# Roger Boscovich
## The founder of modern science

Dragoslav Stoiljkovich

Translated and edited by
Roger Anderton

Petnica Science Center
2010 (Serbian edition)

LULU
2014 (English edition)

Translated from Serbian edition:

Petničke sveske broj (Petnica's Papers No.) 65
ISBN 978-86-7861-043-1
ISSN 0354-1428
Urednik (Editor): Branislav Savić

RUĐER BOŠKOVIĆ - UTEMELJIVAČ SAVREMENE NAUKE
Autor (Author): Dragoslav Stoiljković

Recenzenti (Reviewers):
Dr Slobodan Jovanović
Dr Nikolai Ostrovski
Dr Radmila Radičević
Dr Radoslav Dimitrić
Dr Aleksandar Tomić

Lektura i korektura (Proofreading): Dušica Božović
Direktor (Director): Vigor Mojić

Štampa (Printed by): Valjevoprint, Valjevo

Izdavač (Publisher): Istraživačka stanica Petnica (Petnica Science Center)

Valjevo, 2010.

# CONTENT

| | |
|---|---|
| FOREWORD OF ENGLISH EDITION | I |
| FOREWORD OF SERBIAN EDITION | I |
| 1. LIFE OF ROGER BOSCOVICH | 1-1 |
| 2. ACTIVITIES OF ROGER BOSCOVICH | 2-1 |
| 3. BOSCOVICH'S "THEORY OF NATURAL PHILOSOPHY" | 3-1 |
|    3.1. A unique law of forces that exist in nature | 3-1 |
|    3.2. Orbitals in Boscovich's Theory | 3-2 |
|    3.3. Quantum meaning of Boscovich's Theory | 3-3 |
| 4. CONTRIBUTION OF BOSCOVICH'S THEORY TO MODERN COMPREHENSION OF THE STRUCTURE OF MATTER | 4-1 |
|    4.1. Common view of the historical journey for the discovery of structure of atoms, molecules and macromolecules | 4-1 |
|    4.2. Contribution of Boscovich's Theory to the discovery of the structure of atoms | 4-2 |
|    4.3. Boscovich's comprehensions of elementary points, atoms and molecules | 4-5 |
|    4.4. Macromolecular hypothesis of Boscovich | 4-9 |
|    4.5. Nano-tubes, diamond and graphite | 4-10 |
|    4.6. Boscovich's signposts to neutrino, gluons and quarks | 4-11 |
| 5. CONFIRMATIONS OF BOSCOVICH'S FORCE LAW IN MODERN SCIENCE | 5-1 |
|    5.1. Relation of force and energy dependence on the distance between particles | 5-1 |
|    5.2. Interaction of atoms | 5-2 |
|    5.3. Interaction of molecules | 5-3 |
|    5.4. Interaction of nano-particles | 5-5 |
|    5.5. Interaction of macromolecules | 5-5 |
|    5.6. Interaction of colloidal particles | 5-6 |
|    5.7. Fission of heavy atomic nuclei | 5-7 |
|    5.8. Energy of atomic nucleus | 5-8 |
|    5.9. Interaction between nucleons and $\Lambda^\circ$ hyperon | 5-9 |
|    5.10. Conclusion concerning the validity of Boscovich's curve | 5-9 |

## 6. COMPRESSION OF MATTER - REFLECTIONS OF BOSCOVICH'S THEORY IN SAVICH-KASHANIN THEORY — 6-1
6.1. Introduction — 6-1
6.2. Material density changes according to Boscovich's opinion — 6-1
6.3. Material density changes according to Savich-Kashanin theory — 6-2
6.4. Relation between Boscovich's opinion and opinion of Savich and Kashanin — 6-3
6.5. Density change in the compression of matter by the model of Savich and Kashanin — 6-5
6.6. Mean densities of planets in Solar system calculated by model of Savich-Kashanin — 6-7
6.7. Adaptation of stepwise mathematical model by actual empirical data — 6-7

## 7. APPLICABILITY OF BOSCOVICH'S THEORY — 7-1
7.1. Introduction — 7-1
7.2. Meaning of critical volumes of matter — 7-1
7.3. Characteristic volumes of matter — 7-3
7.4. Physico-chemical state and polymerization of compressed ethylene gas — 7-6
7.5. Effect of pressure on polyethylene melting point — 7-9
7.6. Structure of fluids based on Boscovich's Theory and Savich-Kashanin's Theory — 7-11
7.7. Polymerization of methyl methacrylate — 7-14

## 8. PHILOSOPHICAL FOUNDATION OF BOSCOVICH COMPREHENSIONS
8.1. Introduction — 8-1
8.2. Attraction and repulsion
– Comprehensions of Boscovich, Hegel and Engels — 8-1
8.2.1. Boscovich's comprehension — 8-2
8.2.2. Hegel's comprehensions — 8-2
8.2.3. Engels' comprehension — 8-3
8.2.4. Distinctions and similarities of Boscovich, Hegel and Engels comprehensions — 8-7
8.2.5. Analysis of Boscovich's comprehension of attractive and repulsive forces — 8-8
8.2.6. Differentiation of matter — 8-10

## 9. ROGER BOSCOVICH – THE FOUNDER OF MODERN SCIENCE — 9-1
9.1. Influence of Boscovich's Theory on the contemporaries and followers — 9-1
9.2. Resurrection of Boscovich's Theory — 9-2

REFERENCES
Note about author

# FOREWORD OF ENGLISH EDITION

The English edition of this monograph is almost identical to the Serbian edition published in 2010 by Petnica Science Center, however, revised, extended and corrected. A new Chapter 7.6. titled "Structure of fluids based on Boscovich's Theory and Savich-Kashanin's theory" was added. The Chapters 7.4. and 7.7. are slightly extended, all in order to clarify the applicability of Boscovich's Theory. Table 8-2 and corresponding explanation are added, too.

The author is very grateful to Roger Anderton for translating and editing this book.

*Author,*
April, 2014

# FOREWORD OF SERBIAN EDITION

It's been two and a half centuries since Roger Boscovich published his monumental work "A Theory of natural philosophy reduced to a one unique law of forces that exist in nature". The Theory has had a major impact on Boscovich's contemporaries, scholars of the 18th century, and this resulted in many followers in the 19th and at the beginning of the 20th century. It was studied in many educational institutions all over the world, and was present in many textbooks, books, encyclopaedias.

And then suddenly everything went silent. Today it is no longer present in the curricula of schools and colleges. Apart from the few individuals, our contemporaries, even highly educated people, know almost nothing about Boscovich. They knew hardly anything about who he was, when he lived and what he did, and why many streets and other institutions carry his name.

In the last century in our (Serbian) language there has been published several monographs and a number of professional articles on Roger Boscovich which shows his life and scientific activity, his scientific and philosophical views, as well as his influence on contemporaries and followers. These topics are dealt with briefly and succinctly in this monograph (Chapters 1 and 2).

However, scholars and interpreters of Boscovich's life and creativity have completely missed some of his concepts or have only partially processed them, and sometimes completely misrepresented them. Therefore, it is the basic purpose of this monograph to shed light on those issues which the existing literature has not devoted enough attention.

It is almost unnoticed that his Theory is actually the first quantum theory. He was the first one to draw the orbitals by which a particle moves around particles located in a centre and explain that by transition from one orbital to another a particle either gains or loses a certain amount (quantum) of energy. Having summarized his Theory in Chapter 3, then we present Boscovich's orbitals and interpret the quantum meaning of his Theory.

Little is popularly known about Boscovich's contribution to the contemporary understanding of the structure of matter. His primary undoubted contribution was to the discovery of the structure of atoms of chemical elements. Boscovich's understanding of elementary points, atoms and molecules are usually incomplete and misinterpreted by scholars. Also the scholars of Boscovich's work have completely missed that he was on his way, with his vocabulary, pointing to the possibility of existence of macromolecules (i.e. polymers) and nano-tubes. He described the structure of these materials and their basic properties, and also the structure of diamond and graphite. Following the signposts (of his line of thought), the ideas of neutrino, quarks and gluons can be reached. These topics are dealt within Chapter 4 of this monograph.

The foundation of Boscovich's Theory is the well-known (to some academics) Boscovich's curve that describes the change in force between the particles of matter depending on the distance between them. It is striking that the investigators of his work have never asked themselves whether the curve is scientifically confirmed. This question was never considered, and hence the answer was not sought. Therefore, in Chapter 5 we list a dozen examples that confirm the validity of Boscovich's curve at several levels in the hierarchy of matter - of nucleons in atomic nucleus to the colloidal particles.

Serbian famous scientists, Savich and Kashanin, in the middle of the last century presented a theory about the behaviour of matter at high pressures. Among other things, it can be applied to calculate the mean density of the planets in the Solar system. Although they do not refer to Boscovich, there is obviously a real connection with his Theory. With some adaptation of the mathematical model, we can obtain a significantly more accurate calculation for the results of planet density (Chapter 6).

The value of Boscovich's Theory is reflected in the multitude of ideas that sprout from it and that can be used to solve some of the problems of modern science. Using his directions we approached an original interpretation of the meaning of a critical point, and gave a mathematical model to calculate the volumes of matter at critical point conditions and in some other characteristic states of matter. His Theory was also a signpost to the interpretation of the mechanism and kinetics of the polymerization of ethylene and methyl methacrylate. This topic is devoted to Chapter 7.

There is a plenty of writings concerning the philosophical views of Boscovich. It is written that, from his perception, the attractive and repulsive forces are the essence of the behaviour of matter. However, it is missed that Kant, Hegel, Engels and many other philosophers had similar views. A comparison of Boscovich's understanding of attractions and repulsions with the understanding of these other philosophers is stated in Chapter 8.

Bearing in mind the influence of Boscovich's Theory on contemporaries and subsequent scholars we derive the conclusion that his Theory directly, and sometimes indirectly, have been built into the foundations of modern science.

Everything we have written in this book has been already published in scientific, professional and philosophical journals and presented in a number of scientific meetings. However, in each of them only some particular issues of Boscovich's life and work are considered. All these considerations are scattered in many national and international journals and conference proceedings. This book is arisen as a need to collect them into one place, to interconnect those ideas and then to give readers a complete view of Boscovich's contributions to modern science. Only then, can it be seen to what extent his Theory has been built into the foundations of contemporary science, and thus reveal how far he was ahead of his time, and according to some beliefs he is ahead of our present time.

Initial Chapters of the book are written in a popular and interesting way to be understood by those whose knowledge of natural science is not their strong point. Other Chapters are slightly more complicated. In order to understand them it requires some knowledge of physics, chemistry and science of polymeric materials.

The author thanks the reviewers, and also, Dr. Matilda Lazich, for very useful suggestions and comments, which have contributed to the quality of the monograph. The author is also grateful to the Petnica Science Center, which accepted this monograph for publication.

*Author*,
2010

# Roger Boscovich
## The founder of modern science

# 1. LIFE OF ROGER BOSCOVICH

In the past decade there has been published several books /1-7/ which describe in great detail the life of Roger Boscovich. The reader is referred to the literature cited, and here we will only briefly point out the main themes of his life that are important for the creation and development of his scientific and philosophical thoughts.

Boscovich was born on May 18th, 1711 in Dubrovnik, as the eighth of nine children of a father Nikola and mother Pavla. Nikola was born in the Herzegovinian village Orahov Do, and the mother of Pavla came from a family of Betera in Bergamo in northern Italy. Roger attended the Jesuit College of Dubrovnik where his exceptional talent was observed. To continue his education, in 1725 he went to the Roman Jesuit College where he studied rhetoric, logic, philosophy, mathematics, astronomy and theology. In 1733, he became a teacher of grammar, and in 1740 he was professor of mathematics at the Roman College. In Rome he remained until in 1759 during which period he developed his scientific and philosophical concepts. He published a number of papers among those which are significant are "De viribus vivis" ("The Living Forces" in 1745) and "De Lumine" ("The Light", 1748), along with many works on astronomy, and his most important work "Philosophiae naturalis theoria redacta ad unicam legem virium existentium in natura" ("A Theory of natural philosophy reduced to a one unique law of forces that exist in nature") /8/. The first edition of the Theory was printed in Vienna in 1758, and the second edition 1763 in Venice, which was then in Austria. (*Editor's note: There have been many boundary changes of countries over the centuries.*)

Boscovich was a member of the Catholic order of the Jesuits (Society of Jesus), whose important role was to defend the church from heretical teachings. At that time the church taught that the Earth was the centre of the universe and that the sun and planets revolve around it (geocentrism). On the list of banned books were heliocentric teachings, according to which the Earth and the other planets revolve around the sun. Although it was already in full development (in England), on the list of undesirable teachings was also Newtonian mechanics. The part of Newtonian mechanics, which refers to the movement of the planets around the sun, was banned by the church.

Although a Jesuit, Boscovich had convinced himself to be a Newtonian. He advocated the acceptance of Newton's teaching, but also feared that he might come under attack by the Jesuits, to which he belonged. His Theory was therefore published in Austria, where there was less Jesuit influence. He realized, however, that too conservative a point-of-view ruled in the Roman College and that this was not the environment to keep pace with the latest scientific and philosophical achievements, thus Rome was not the place to develop and present such views.

Therefore, in 1759, he went on a long study trip to Europe, without any desire to return to Rome. He went first to Paris, where as a corresponding member of the Royal Academy of Sciences (admitted on May 6th, 1748) he attended meetings of the Academy. There he met many famous encyclopaedists, and became introduced to their ideas. He went to London in 1760, where on January 15th, 1761, he became a member of The Royal Society; the Royal Society represents English academy of sciences. After England, from about 1760 to 1763, he visited many well-known scientific and public institutions and individuals in the areas that now form parts of Netherlands, Belgium, Germany, Austria, Turkey, Bulgaria, Moldova and Poland.

Upon completion of his travels, he refused to go back to the Roman College, but in 1764 accepted professorship of mathematics in the small town of Pavia near Milan, which was then within Austria. In 1765, he accepted the invitation to establish an astronomical observatory in Milan, together with the Jesuits from Brera. Brera is a district that still exists in the centre of Milan (Figure 1-1). All his intellectual forces and financial funds he invested in the construction of the observatory (Figure 1-2), and in 1770, he moved to Milan, to be professor of astronomy and optics. Until in 1772, he was the head of the observatory, together with Abbot Le Grange.

*Figure 1-1. The entrance to the Brera (Milan, recorded 2003)*

*Figure 1-2. Photo models of the Boscovich observatory*

In Brera there is still the Astronomical Institute founded by Boscovich. The name of the founder Roger Boscovich is still prominent on the board at the entrance of the Institute (Figure 1-3). In the lobby of the Institute, there is a bust of Boscovich (Figure 1-4), as well as the telescope from the period of his work in Brera. On the roof of Brera, in a place where once there was Boscovich's observatory, today is a modern observatory (Figure 1-5).

*Figure 1-3. The board at the entrance of the Astronomical Institute in Brera, which reads Boscovich as founder of the Institute*

*Figure 1-4. Bust of Boscovich at Astronomical Institute in Brera*

*Figure 1-5. Modern observatory in Brera is located where once was Boscovich's Observatory*

Due to disagreements with Abbot Le Grange as well as with other Jesuits in Brera, Boscovich left Milan in 1772. He was then an elderly man of 61 years, left with no income and no savings after using them to establish the observatory. Unfortunately for Boscovich, in 1773 the Jesuit order was abolished so that even from them he could not expect help. Therefore in 1773, he accepted the offer of the Ministry of the Navy of France to be their director of the Department of optics. He moved to Paris and took French citizenship.

To prepare for printing his works on optics, in 1782 he obtained leave to reside in Milan, where in 1785 his book was published "Opera pertinentia ad opticam et astronomiam" ("Acts relating to optics and astronomy"). Exhausted from work, and mentally ill, Roger Boscovich died in Milan on February 13th, 1787.

He was buried in the church of St. Mary Podone, which is located on the piazza Boromeo in down-town of Milan (Figure 1-6). At the entrance to the church is a relief of Boscovich (Figure 1-7) where still can be read the letters "...CO ...VI..." as part of his last name, written in Latin. Boscovich's body was placed in a recess in the wall of the church. The church was unfortunately destroyed in the bombing during World War II /1, p. 1041/. After its restoration, Boscovich's burial place was sealed off. At the Milan cemetery in the Pantheon called "Famedio" there is a memorial plaque with only Boscovich's name and the years 1711 - 1787.

*Figure 1-6. Church of St. Mary Podone in the centre of Milan where Roger Boscovich was buried (recorded 2011)*

*Figure 1-7. Relief of Boscovich at the entrance to the church of St.Mary Podone (recorded 2011)*

Boscovich was a celebrity in his time, a member of several academies of sciences (English, French, Russian...), renowned as an astronomer, physicist, mathematician and philosopher. The impressiveness of the versatile activities of this great scientist is perpetuated in the "heart" of his scientific activity, in magnificent Milan. One street in the centre of Milan even bears Boscovich's name (Figure 1-8) with a memorial plaque dedicated to him (Figure 1-9).

*Figure 1-8. Street in Milan, named after Roger Boscovich*

*Figure 1-9. Memorial plaque in the street of Roger Boscovich in Milan*

# 2. ACTIVITIES OF ROGER BOSCOVICH

Boscovich's versatile activities were in many different scientific fields. Here in this chapter we will only briefly outline his scientific and philosophical activities. A more detailed account of his life and work can be found in the books of domestic /1-5, 9/ and foreign authors /6, 7/. The books by Markovich /1/, Dimitrich /4/ and White /6/ give lists of more than one hundred works by Boscovich.

Basic scientific and philosophical concepts, which Boscovich dealt with include: continuity and discontinuity of matter, space, time and motion /10, 11/; issues of divisibility and combinability of particles of matter, the forces that govern between these particles, the nature and use of infinitely large and infinitely small sizes. Although to these issues he devoted specific works, their unified representation, as well as their elaboration and implementation are given in the most important of Boscovich's work "Philosophiae naturalis theoria redacta ad unicam legem virium existentium in natura" ("The theory of natural philosophy reduced to one unique law of forces that exist in nature") /8/. In that work the basic issues of the structure of matter are considered: starting from elementary points, via the atoms, molecules, macromolecules and so up to the celestial bodies. (A more detailed presentation of the Theory is given in Chapter 3. of this book.) Boscovich's Theory suggests a unique law of forces between the particles, as follows: at the large distance between the particles there is an attractive (gravitational) force, and with decreasing distance a repulsive force arises, then again becomes attractive... And so on several times until at small distances there arises a large repulsive force that prevents contact of the particles (or bodies). The unique law of forces Boscovich displayed as an oscillating "force-distance" curve in which the attractive and repulsive arches turn alternately (so called Boscovich's curve, Figure 3-1). This law of forces, Boscovich applied to the interpretation of various phenomena in physics, mechanics, optics, chemistry and astronomy, which were known in his time. (Confirmation and implementation of Boscovich's law of forces in modern science are presented in Chapter 5 to 7 of this book).

A large part of Boscovich activities was related to many theoretical and practical issues in the field of astronomy. He elaborated theoretically the construction of astronomical instruments and assessed their reliability. He constructed a ring micrometer and achromatic telescope.

Applying the theory of gravity, he considered the movements of bodies in the Solar system and what emerges as consequences of these movements (tidal sea, the shape and structure of the Earth).

In 1760, he published in London, in the form of a poem, a work on the eclipses of the sun and the moon, which was reprinted in 1761 in Venice and in 1767 in Rome; in 1779 it was printed in Paris in the French language. Recently, the poem, which is about 300 pages with 1550 verses, was published in Serbian /12/. Contrary to Euler, Boscovich suggested that the moon has no atmosphere. While using data from shifts in sunspots, and applying his own method, Boscovich determined the rotation time of the sun.

He developed an original method for determining the orbits of the planets and comets. When the English astronomer Herschel in 1781 discovered a new celestial body in the Solar system it was initially thought to be a comet. Boscovich, it could be said, was among the first who accurately determined the path of the body and concluded that it was not a comet but a new planet, which is now known as Uranus /1, 2/.

As a great scientist he paid special attention to the theoretical interpretation and application of optical issues: nature of light, its propagation, diffraction and scattering, improving optical instruments.

In mathematics Boscovich made major contributions. He hollowed out (i.e. defined and explained) many mathematical concepts and provided original solutions of mathematical problems, primarily in the area of geometry. He also developed spherical trigonometry. It is also interesting to point out that Boscovich first developed a theory for analysis of measurement errors (long before Gauss and different from Gauss) /4/.

There are reports of Boscovich's contributions in the field of engineering, civil engineering, architecture, hydraulic engineering, as well as contributions to archeology.

It is also worth mentioning his famous diary of his journey from Constantinople to Poland in 1762. This was the "literary and scientific work of the first order, with French, German and Polish translations occurring before the Italian originals and were snapped up by readers so fast that even Boscovich could not get copies." (Preface of D. Nedeljkovich the Serbian edition of the diary /13/).

# 3. BOSCOVICH'S "THEORY OF NATURAL PHILOSOPHY"

## 3.1. A unique law of forces that exist in nature

Boscovich's comprehension of nature is partly based on the ideas of Leibniz and Newton, and partly deviates from them. From Leibniz he accepts the assumption that the basic elements of matter are tiny dots (monads), which do not have size (non-extended) and are indivisible. However, Boscovich does not accept Leibniz's assumption that the points can touch each other. Instead, he believes that the points are distanced from each other over some space, which can infinitely increase or decrease, but not completely disappear. From Newton, he accepts the existence of mutual forces between these points. Unlike Newton, who believes that at very short distances there is strong attractive force between the particles, Boscovich believes that there is a strong repulsive force, which becomes greater as the distance becomes less. This is similar to the views of Empedocles that there are forces of love and forces of strife, where Boscovich believes that force can be attractive or repulsive, and which alternates depending on the distance between points (Figure 3-1).

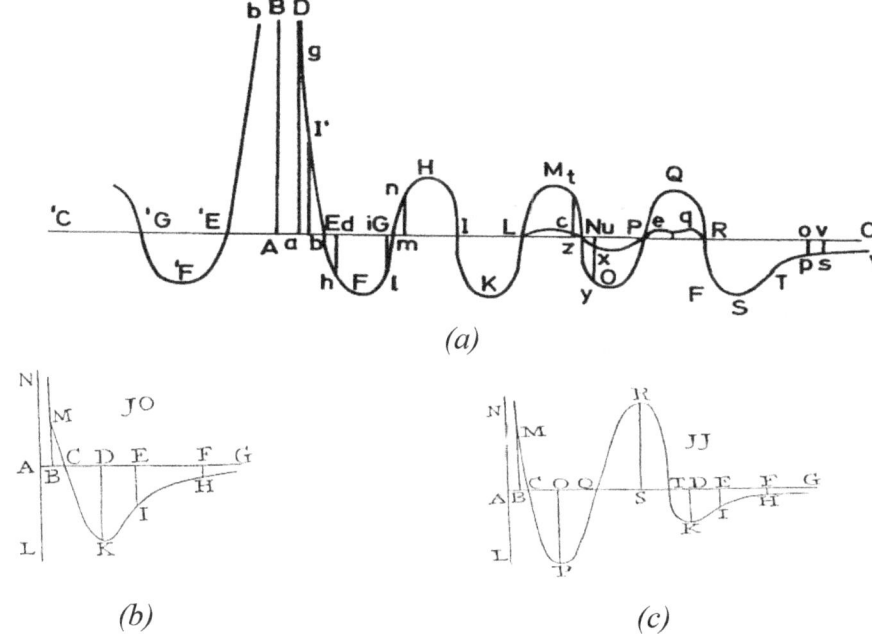

*Figure 3-1. General (a) and special shapes of Boscovich curve show the change of attractive and repulsive forces (upper and lower ordinate, respectively) with the change of the distance (abscissa) between the elementary points of matter /8, Figure 1/*

Boscovich accepted Newton's notion that assembling points together in collections forms more complicated particles of the first order, connecting these first order particles together then forms particles of the second order, then the third order and so on, and by further assembling atoms are formed which are not themselves elementary particles but consist of parts. He considered that molecules are even larger assemblies of particles. Moreover, Boscovich in 1758 was the first to put forward the suggestion that the macromolecules could exist as series of atoms, and he describes their spiral structure and properties! However, today it is commonly said that the macromolecular hypothesis was first proposed by Staudinger in 1920!? (This is described in more detail in Chapter 4.)

According to Boscovich, there are individual stages (ranks) in the hierarchy of matter, i.e. elementary points, then particles of first order, then second order, atoms, molecules, and even the entire Solar system. Boscovich indicates that "all worlds of smaller dimensions, taken together, were like a single point in relation to the larger" world. (The forces acting among the lower orders of the particles are larger than the forces between the particles of higher orders /8, Section 424/.) He believes that his curves shown in Figure 3-1 are valid for each pair of particles at any level of the hierarchy of matter. The number of arches, their size and shape, however, can be different: curve with one attractive and repulsive arch (Figure 3-1b), two attractive and repulsive two arches (Fig. 3-1c), as well as with a number of arches (Fig. 3-1a).

Boscovich especially points out that there are distances (E, G, I, L, N, P, and R in Figure 3-1a) at which the repulsive and attractive force are equal. The particles are in balance if they are at such distances. However, there are two types of cases. In some cases (E, I, N and R) by increasing the distance the attractive force increases, and by reducing the distance the repulsive force increases. In this case, the particles are in a stable equilibrium, for if the distance between the particles is accidentally increased, then it creates an attractive force, and brings them back to the previous distance. If, however, the distance is reduced, then the resulting repulsive force brings them back to the previous distance. These distances (from A to E, I, N and R) are called **the limits of cohesion.**

In other cases, if the distances are corresponding to the positions G, L and P, the particles are in an unstable equilibrium because you either have (a) a small increase leading to the appearance of the repulsive force and to an even greater separation of the particles, or (b) a small decrease in distance leading to the appearance of the attractive force and to an even greater approach of the particles. These positions Boscovich called **the limits of non-cohesion.**

## 3.2. Orbitals in Boscovich's Theory

When considering the mutual effects of three points, Boscovich indicates a "beautiful theory about the point placed on the ellipse, while the other two points occupy the foci of the ellipse" /8, Sections 230-235/. Specifically, if the two points (or particles) located at focal points A and B near the centre of D, then the third point may be located at anywhere on the ellipse at a distance corresponding to the limits of cohesion (Figure 3-2). Accordingly, there are as many ellipses as there are cohesion limits.

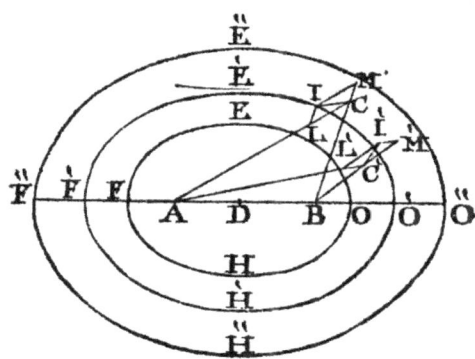

*Figure 3-2. Orbitals in Boscovich's Theory /8, Figure 33/*

## 3.3. Quantum meaning of Boscovich's Theory /14, 15/

Boscovich indicates that "in mechanics it is known that for a curve, whose abscissas represent distances and ordinates represent forces, then the area (delimited by the curve and abscissa) represents the increase or decrease of the square of velocity" of the particles that move mutually /8, Sections 118, 176, 191/. (In this book, Chapter 5-1 in Figure 5 it is shown that $F = -dE/dr$; wherein: force F, distance r, and the kinetic energy E of a particle of mass m moving at a speed v is $E = mv^2/2$. Hence it follows that the area below or above the arches of the abscissa in Figure 3-1 represent energy.). Therefore, the individual areas, delimited by abscissa and (repulsive and attractive) arches, are a measure of the increase or decrease of the square of speed of the particles as they approach (or separate from) each other. The two particles upon approaching will remain at a distance apart, the size of which depends on their initial rates and the area delimited by attractive and repulsive arches. Boscovich explains that if the areas delimited by repulsive arches are less than the attractive areas, the particles will reach the first limit of cohesion (position E in Figure 3-1) by a speed that is proportional to the surface delimited by the first attractive arch (EFG) and moving on a circle having radius of AE will continuously oscillate around that limit.

Moreover, Boscovich indicates that it stems from his Theory that as a particle approaches another particle, and when it passes from one to the other limits of cohesion, it will lose or gain exactly a certain amount of energy. That "quantum energy", as it is now called, between the two limits of cohesion is equal to the difference between areas delimited by repulsive and attractive arches.

Hence, Boscovich's Theory is actually the very first quantum theory. He described what a century later was merely assumed by Planck. Reviewing the quite extensive literature on Roger Boscovich available to us, we find that no other authors (except A. Tomic /16/) observed that Boscovich's Theory is indeed a quantum theory and that Boscovich laid the foundation stone for the discovery and development of $20^{th}$ century quantum theory, which followed a century and a half later. Unfortunately, even the originators (Planck, Einstein and Bohr) of $20^{th}$ century quantum theory and the latter scholars seemed mostly unaware of Boscovich's foundation.

# 4. CONTRIBUTION OF BOSCOVICH'S THEORY TO MODERN COMPREHENSION OF THE STRUCTURE OF MATTER

## 4.1. Common view of the historical journey for the discovery of structure of atoms, molecules and macromolecules

It is often said that the ancient Greek philosophers Leucippus and Democritus first came to idea that all was made of atoms, tiny indivisible particles. Their thought was religiously prohibited and dormant for more than 1500 years. During this period, there were a few people who thought about the atomic structure, but in the period to 19th century, there was a great preparation that formed the basis for further work on it. It is often said, that John Dalton at the beginning of the 19th century came up with the idea that each chemical element has its smallest particles. Believing that these particles are indivisible, Dalton, following the example of the Greeks, called them atoms (Scheme 4-1.).

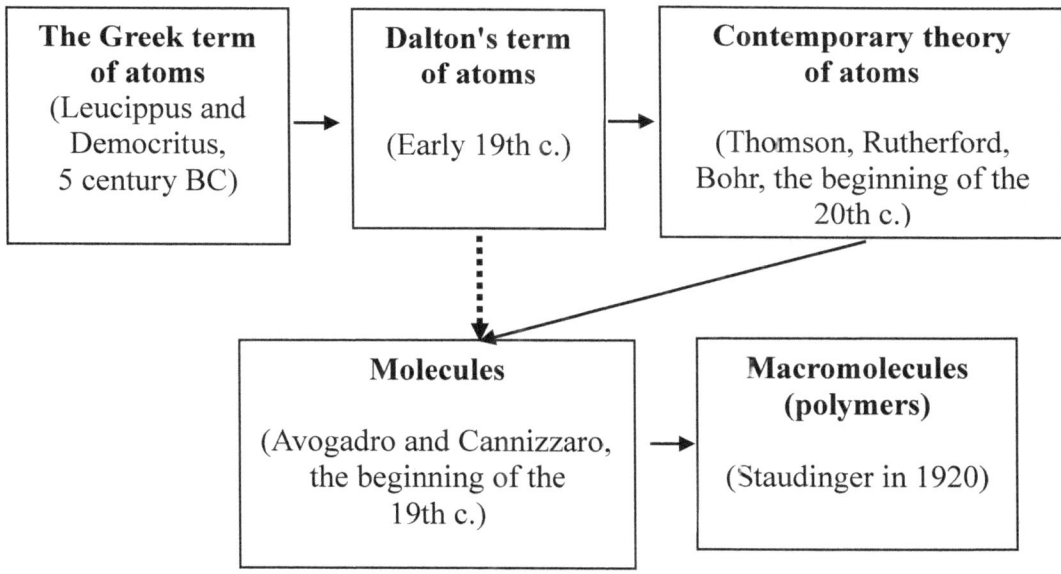

*Scheme 4-1. A common view of the historical journey of discovery structure of atoms, molecules and macromolecules*

A little later it turned out that these Dalton atoms must be divisible, i.e. the atom had a structure, and the atom was made up of smaller particles, the atomic nucleus and electrons. This truth of atoms was revealed in the 19th and 20th century and many famous scientists contributed to finding out the structure of atoms are usually named as: Faraday, Maxwell, William Thomson (better known as Lord Kelvin), J. J. Thomson, Rutherford and Bohr. The remarkable contributions of the last three scientists are emphasized; according to the usual contemporary story for the historical journey of the discovery of the atomic structure looks like shown in scheme 4-1. Then, usually are listed the names of A. Avogadro and A. Cannizzaro who in 19th century indicated that atoms are combined into molecules, and then its stated that H. Staudinger in 1920 first introduced the hypothesis that the molecules combine into even larger entities - macromolecules.

However, it was not quite so. A part of the story was left out. It is undeniable that these scientists contributed highly to the interpretation of the structure of matter. It is important to note, however, that these achievements are based on the ideas of Roger Boscovich, which is not known enough to the wider scientific community.

Earlier in western literature it was regularly cited the importance of Boscovich to the discovery of the structure of atoms, but since 1920, his name is usually omitted /6a/. It is commendable that some of our scholars in Serbia and Croatia typically cite the name of this great scientist, but unfortunately do not give enough information on his impact on the discovery of the structure of atoms. Therefore, we would like here to briefly introduce the reader to the contribution of Boscovich to the discovery of atomic structures, and more detailed views can be found in the literature /2, 6a, 7, 15, 17, 18/.

## 4.2. Contribution of Boscovich's Theory to the discovery of the structure of atoms

At the end of the 19th century, the more mature conviction (i.e. point-of-view) was that Dalton's atoms of chemical elements were still divisible and consisted of positively charged particles and negatively charged electrons. The question was - how were these particles located in the atom.

At the end of the 19th century, J. J. Thomson (from Cavendish Laboratory in Cambridge) discussed various models of atoms. According to one of them, which is most frequently cited in contemporary literature as by Thomson, is that of positive charge filling the entire atom forming a ball, where negative electrons are deployed like plum grains in pudding. (Hence, it is named "plum-pudding model" as well as "Thomson model".) However, Lord Kelvin, in the period 1902-1907, published several works which emphasized his belief that the issue of atomic structure can be resolved by Boscovich's Theory and proposed a "planetary model of the atom".

J. J. Thomson also thoroughly discussed the "planetary model of the atom", under which the positive charge is located in the nucleus of atom and the electrons orbit the nucleus /2, 7/. Seeking a theoretical foundation for the idea that electrons can move only at certain paths around the nucleus of atoms, Thomson concluded that for this purpose only Boscovich's Theory would serve. In 1907 Thomson wrote in his work "The corpuscular theory of matter" /90/: "Suppose we regard the charged ion as a Boscovichian atom exerting a central force on a corpuscle which changes from repulsion to attraction and from attraction to repulsion several times... such a force, for example, as is represented graphically in Figure 4-1 where the abscissa represent distances from the atom, and the ordinates the forces exerted by the atom on a corpuscle..." It is obvious that Figure 4-1 actually combines Boscovich's curve (Fig. 3-1) and Boscovich's orbitals (Fig. 3-2).

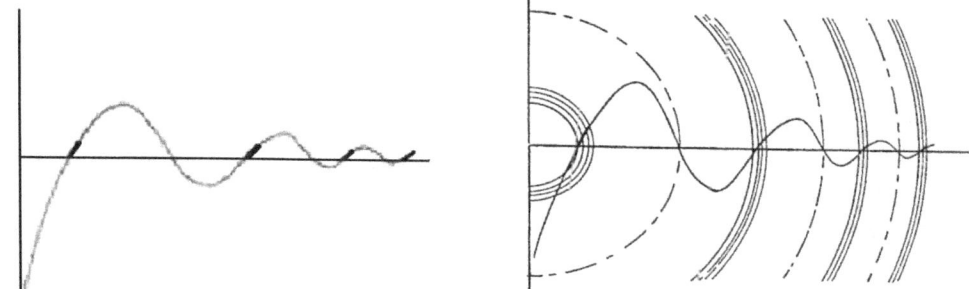

*Figure 4-1. Left curve as stated by Thomson /90/: A positively charged nucleus of the atom is at coordinate's origin and the positions of electron orbits are at bolded part of the curve. Following Thomson's opinion, Gill /7/ presented "permissible" (solid line) and "forbidden" (dashed line) orbitals (right curve). The abscissa shows the distance of the electron from nucleus and the ordinates show the force: repulsive (below) and attractive (above) /7/.*

The doubt, over what model of the atom was correct, the "plum-pudding" or "planetary", was solved by Rutherford, who was a former student and collaborator to Thomson. Rutherford in 1907 transferred to the Department of Physics, University of Manchester, and in the next year confirmed that alpha particles are actually helium nuclei, i.e. positively charged particles which are composed of two protons and two neutrons. Thin sheets of metal were bombarded with alpha particles, and thus Rutherford in 1911 experimentally confirmed the "planetary model of the atom". This model is commonly called a "Rutherford model".

In 1912, after seven months spent with Thomson in Cambridge and four months spent with Rutherford in Manchester /19/, Niels Bohr in 1913 calculated the possible paths of electrons, taking into account that electrons can move from one orbital to another only if they receive or lose a certain amount of (quantum) energy - as Boscovich said a century and a half earlier (Section 3.3). Today, this model of the atom is called "Bohr model", which is not fully justified to call it that.

During the celebration of the two hundredth anniversary of the publication of Boscovich's Theory, held in Dubrovnik, Niels Bohr wrote: "Boscovich's ideas exerted a deep influence on the work of the next following generation of physicists... Our esteem for the purposefulness of Boscovich's great scientific work, and the inspiration behind it, increases the more as we realize the extent to which it served to pave the way for the later developments" /5, p. 8 and 184/.

Therefore Supek /5, p. 184/ rightly asks whether Niels Bohr was really referring to himself when he emphasized the impact of Boscovich on the next generation of physicists. "When I repeatedly talked to him, the answer was never clear. Perhaps such clarity was not in the father of modern physics. If then he did not know Roger's work, he had to have known the Thomson atom which probably makes the influence indirect."

Therefore, bearing in mind that in the period 1903-1907 "J. J. Thomson deducted his hypothesis directly from the Theory and curve of Boscovich, and showed that the notion of 'allowed' and 'forbidden' orbits follows from it", Gill /7/ points out that Boscovich made an "essential element of the modern concept of the atom" and "where Boscovich planted two hundred years ago others have reaped." Hence, Gill called this model "The Boscovich-Thomson" atom and indicates that "when the history of atomic theory is being written, it is right that the part played by Father Roger Boscovich should not be overlooked".

Taking into account the contribution of Boscovich, the actual history of the discovery of atomic structure is shown in Scheme 4-2.

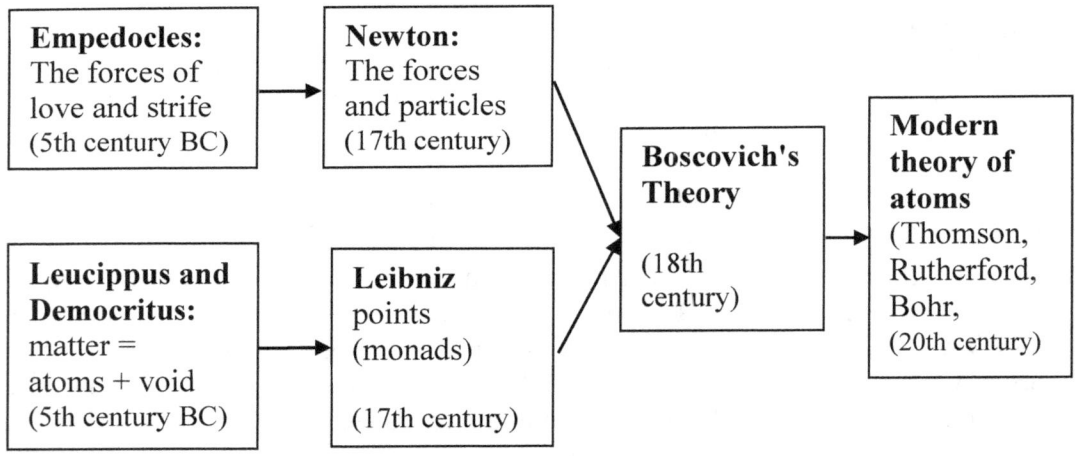

*Scheme 4-2. The actual historical path of discovery of the structure of atoms*

## 4.3. Boscovich's comprehension of elementary points, atoms and molecules

In many descriptions of Boscovich's Theory his comprehension of elementary points are incompletely, and even wrongly interpreted, which leads to the fact that his views in terms of atoms and molecules can be completely ignored. To properly understand his comprehension of elementary points, atoms and molecules, it is first necessary to clarify some basic contemporary concepts in terms of matter and atoms.

In many scientific and popular articles /20/ it is often said that today we can be absolutely sure that matter is composed of atoms. Afterwards it is stated that Greek philosophers Leucippus and Democritus came to this idea in 5th century BC.

However, the modern understanding of the term "atom" and the term "mater" is essentially different from the understanding of Leucippus and Democritus. For these philosophers, atoms are the smallest particles of matter, which have no parts, and thus they cannot to be further divided. The word "atom" comes from the Greek word meaning "indivisible". In fact, according to the ancient Greeks there are many types of atoms, which are different in size and shape.

The contemporary understanding of the term "atom" is more recent and comes from an English scientist Dalton who early 19th century showed that the chemical elements are composed of tiny particles that enter into mutual chemical reaction. Dalton believed that these particles did not have parts and he named them "atoms", like the Greeks did. However, Dalton was wrong in terms of their indivisibility, since in the late 19th century it became known that these chemical atoms still have parts inside them, as it was confirmed in the 20th century. The modern concept of the atom means that they are composed of identical electrons (negatively charged), which range around the atomic nucleus, and the nucleus of an atom consists of identical protons (positively charged) and identical neutrons (not charged). Moreover, it is well known that the protons and neutrons also consist of still smaller particles.

The ancient Greeks would never have called the smallest particles of chemical elements as "atoms", because they are divisible and consisting of smaller particles, but modern science does call them "atoms". Therefore, when the scientific and philosophical works speak of "atoms" and "atomistic" it should clearly be distinguished if what is meant is as per what the Greeks referred to atoms (as the smallest indivisible particles), or the modern concept of atoms, as complex and divisible particles.

In addition, there are misconceptions when describing Leucippus and Democritus in their comprehension of the concept of matter: "Well, of atoms, these tiny particles, it was all done by Democritus" – it was recently stated in an article /20/. It is customary to say that the ancient Greeks believed that matter is composed of atoms. But something like that they have never claimed! In fact, Leucippus and his associate Democritus held the belief that the elements are composed of **fullness and the void**; they call them being and not-being, respectively. Being is full and solid; not-being is void and rarefied. Since the void exists no less than does the body, it follows that not-being exists no less than being. **The two together are the material causes of existing things** /21/. When Democritus says atoms touch, the touch is called the mutual proximity of atoms over short distance, because they certainly can be dismantled. In other words, in their view it is not possible for atoms to touch even in a collision, because there must remain at least a small gap between them.

So, they thought that the matter is inseparable unity of fullness and VOID (emptiness). We repeat: the VOID! According to them, the atoms are only one part of matter (fullness), and the second part is the void (emptiness). The atoms + void, only taken both together, constitute matter.

This is reflected in the structure of chemical atoms, as we know it today. No matter that it is a very small particle, the atom is "a nearly infinite nothingness with little substance concentrated in the nucleus and even less in the electrons swarming around far, far away on the horizon... Imagine a few grains of sand in the middle of a football playing field and several smaller grains that are rushing around the playground - it is [analogous to] an atom increased several times. An emptiness with a few grains, but imbued with forces" /20/.

It can also be easily imagined that in the space from the centre to the rim of the field can accommodate more tons and tons of sand - because the space on the field is really almost empty. (It was calculated that atomic nucleus and electrons occupy only 0.000000000001 % of total volume of an atom, and the residual 99.999999999999 % is void.)

However, despite the relatively large distance between the nucleus of the atom and the surrounding electrons, there cannot be placed even a single electron. Nothing can be placed there. Is it empty space, when nothing else can be settled? Moreover, as long as they are together, the nucleus and the electrons do not leave room in the space between them for anything else. It is possible only to place the electrons in orbitals at the perimeter of the atom, and the same orbital can accommodate a maximum of two electrons only if they synchronize their movements so that their spins are opposite (anti-parallel). And when that happens the orbital is completely filled, nothing more can be located in it /22/. [Translator's note: Only 2 electrons with anti parallel spins can be placed in some orbital, whatever its energy level. However, level 1 can accommodate 2 electrons since it has only one 1S orbital; level 2 can accommodate 8 electrons since it has four orbitals: 2S, 2Px, 2Py, 2Pz; etc.]

Nucleus, electrons, space between them and the space that is formed in the orbitals are one entity, which is now called the atom. If the atom is moved it is as a whole, its inner space is moved, too. Whither it is - there is its inner space, too.

But, to be able to move an atom, you need a space between the atoms. Even the ancient Greeks knew that. Also the outer space between atoms is a prerequisite of their independent existence. And that space is an integral part of the matter in which the atoms are moving independently. Matter at the atomic level consists of atoms and the space in and around them.

The extent to which an atom jealously guards its inner space is shown by the fact that it does not give it up even if it is forced to be joined with other atoms. Both atoms will give off some external electrons, and its outer orbitals transform into common inter-atomic or molecular orbitals. However, the inner orbitals and the space in-between the nuclei and electrons in each of the connected atoms will be preserved, perhaps slightly modified.

And then the story gets repeated, but this time at the molecular level. Again, it can be seen that the molecules does not consist only of atoms, but also consist of the space between atoms and within atoms, and matter at the molecular level must also include the intermolecular space.

Therefore, when we ask "What is all this is made off?" we should keep in mind not only the particles, i.e. electrons, protons, neutrons, nuclei, atoms, molecules... Since that is just a part of the matter, namely its "fullness". But the matter is not only the "fullness", but the void (emptiness), too. Thus the real answer requires that for each particle we take into account the space in it and around it.

Consider now Boscovich's comprehension of the structure of matter. According to him, the smallest parts of matter are elementary points, which are indivisible and without size, i.e. non-extended. All of these points are identical; they do not differ for each other. Boscovich said that the idea of non-extended and indivisible elementary points was taken from Leibniz. These are actually the monads of Leibniz. But unlike Leibniz, who held that monads touch each other and therefore matter is continuous, Boscovich believed that the points can not touch, and one point interacts with another by the attractive and repulsive forces according to the law presented in Figure 3-1. Between two points there must always exist at least some very small space.

Therefore, Boscovich's elementary points are different from the concept of atoms of the Greek philosophers, and also from the contemporary understanding of the concept of the chemical atoms (Table 4-1). Hence, it is wrong that some authors call Boscovich's elementary points as "Boscovich's atoms". Boscovich never called his points as atoms.

Table 4-1. The main features of ancient Greek and modern atoms and Boscovich's elementary points

|  | Divisibility | Shape and size | Versatility | Possibility of combining |
|---|---|---|---|---|
| Atoms of Leucippus and Democritus | Indivisible | Different shape and size | Various | Combine without contact |
| Boscovich's elementary points | Indivisible | Non-extended | Identical | Combine without contact |
| Contemporary understanding of atoms | Divisible | Different shape and size | Various | Combine without contact |

However, Boscovich in his Theory also considers atoms /8, Section 440/. The term "atom" Boscovich implies for a particle that is composed of parts, and these parts remain together in an atom owing to the force described by his curve. It should be noted that Boscovich indicates that the atoms have parts, a half century before Dalton!

By Boscovich, atoms are combined into larger particles. "In the case of two particles of which one has approached the other with a very great velocity, there arises a fresh connection of great strength, that is, one so strong that there is no rebound of the particles from one another. For instance, it may be said that the hook of the one is introduced into an opening in the other..."/8, Section 440/. If he had that "hook" of an atom named as electron, and the "hole" of the other atom as an incomplete filled atomic orbital, then it would fully correspond to the modern interpretation of the chemical bond.

Boscovich also uses the term "molecule" and suggests that it can be seen by microscope /8, Section 188/. (We now know that the particles that modern science calls "molecules" can not be seen by optical microscopes, which existed in Boscovich's time.) It is important to note that the particles that Boscovich means by molecule are larger compared to atoms. It is also important to note that Boscovich suggests the existence of molecules, but more than half a century before Avogadro and a century before Cannizzaro, who are usually attributed to the discovery of molecules!

By Boscovich, a molecule is a particle of higher order then atoms. He indicates that the particles of higher orders may be different. First difference comes from the number of points that make up the particle /8, Section 419/; then, because of the different disposition of points /8, section 420/. From these differences in the number and distribution of points the other important differences emerge that influence a large variety of bodies and natural phenomena /8 Section 421 /. This primarily refers to the force that one particle has acting on another. So, there are particles which are attracted, or which are repelled, or which are inert /8, Section 422/. Today it is now known that particles with the same charge repel but with different charges attract one another and also there are uncharged particles that are inert.

Another important difference among the forces of these particles is that one side of some particles are able to attract a second particle, while the other side will repel /8, section 423/.

Boscovich's description of the behaviour of higher order particles is in line with the modern description of the behaviour of molecules. It is now known that many kinds of molecules do not have a uniform distribution of positive and negative charge, due to the fact that some of the atoms in a molecule more strongly attract electrons and some do so weaker. Therefore, the molecules are polarized, i.e. are dipoles - on one side of the molecule is partially positive, and on the other side partially negative charge. It is known that two dipoles mutually repel each other if they approach each other with their sides having the same type of charges. Two dipoles are attracted to each other if one positive charged side of the dipole approaches the negative charged side of another.

## 4.4. Macromolecular hypothesis of Boscovich /23-25/

Today it is a widely accepted view that the German Herman Staudinger was the creator of the hypothesis of the existence of macromolecules (i.e. polymers) and he first presented the hypothesis in 1920. However, this view was not widely accepted before 1930, and there were conflicting opinions in the coming decades.

In this section we want to show that Boscovich was actually the first who announced the possibility of the existence of macromolecules in 1758, in his Theory /8, Section 440/. Bearing in mind his curve, Boscovich suggests that atoms can link (together): "In such a way **atoms might be formed like spirals;** and, if these spirals were compressed by a force, there would be experienced **a very great elastic force** or propensity for **expansion**". Furthermore he stated that "a force being produced at each distance, the figure might suffer some change; and **by a very slight change of each of the distances** in a very **long series of points** there might be obtained a **bending of the figure of comparatively large amount**, due to a **large number of these slight bendings**." In those statements it highlights some of Boscovich's ideas that deserve to be further considered and interpreted in the light of modern concepts in the science of macromolecules.

Boscovich indicates "a long series of points" and "spirals of atoms". That is identical with the modern comprehension about the existence of macromolecules as the chains of chemical bonded atoms. Furthermore, he indicates that these chains could be "very long" and could have "a large number of bendings". Using contemporary scientific words, Boscovich is indicating that there may be a high degree of polymerization. (Under the polymerization degree is meant the number of molecules that are linked to a macromolecular chain.)

By Boscovich, these arrays of atoms may be "spiral" shaped, which represents the polymer chain conformation. In modern science of polymers it is well known that some natural and synthetic polymers do have a spiral (helical) conformation (Figure 4-2).

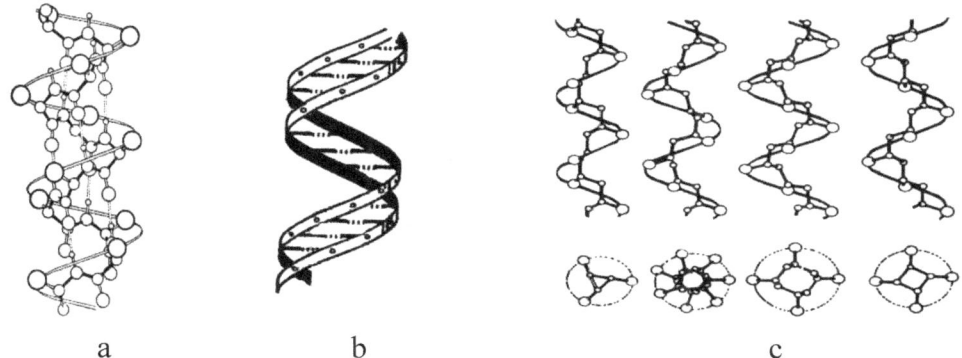

a  b  c

*Figure 4-2. The spiral structure of some natural and synthetic macromolecular chains (a–proteins, b–deoxyribonucleic acids, i.e. DNA, c–polyolefins)*

Boscovich also indicates that "...by a very slight change of each of the distances in a very long series of points there might be obtained a bending of the figure of comparatively large amount, due to a large number of these slight bendings." In contemporary scientific words - conformation of the whole chain can be changed by bending a large number of chemical bonds between the atoms in the chain.

In his statement that these series of atoms can have a huge "elastic force to the expansion" one may recognize a hint of the high elasticity of polymer materials, which is one of the basic features of most polymers.

Obviously, Boscovich pointed to all the basic features of macromolecules: chain structure, a high degree of polymerization, the possibility of helical conformation of the chain, conformation change due to bending of chemical bonds, and even the elastic properties of macromolecular materials. Boscovich suggested this almost two centuries before Staundiger introduced his macromolecular hypothesis.

## 4.5. Nano-tubes, diamond and graphite

Boscovich points out /8, Section 440/ that it is possible that there are atoms whose force curve is as shown in Figure 3-1b. Point C on the curve represents the stable distance of such atoms. There could be "inscribed a continuous series of little cubes, and points are situated at each of their corners". That series would have great persistence in maintaining its shape. Speaking in modern terms, this would be a nano-tube of square cross-section.

If the particles were deployed in the tack of proper pyramid (tetrahedron) and at a distance that corresponds to the limit of cohesion then the body will be "an unbreakable and impermeable solid" with infinite resistance and inflexibility /8, Sections 239, 363 and 419/. But if the particles are not deployed in the tack of the pyramid, or if they are not at the appropriate distances, we cannot then speak of the great strength of the body. If these particles were in one plane, then it would be a flexible material and "could even be folded in spirals after the manner of ancient manuscripts (i.e. rolled scrolls)" /8, Section 362/.

The hardness of diamond and graphite softness is the confirmations of these perceptions by Boscovich. H. Davy (1778-1829) went with the Boscovichian atomistic idea to explain the structure of molecules, different crystal forms, as well as to solve the problem of the structure of the diamond /5, p. 153/. Both diamond and graphite are composed of the same kind of atoms - carbon atoms. But in the diamond these carbon atoms are deployed in the tack of the tetrahedron, while in graphite are in the plane (Figure 4-3), and this is the main cause of their different hardness.

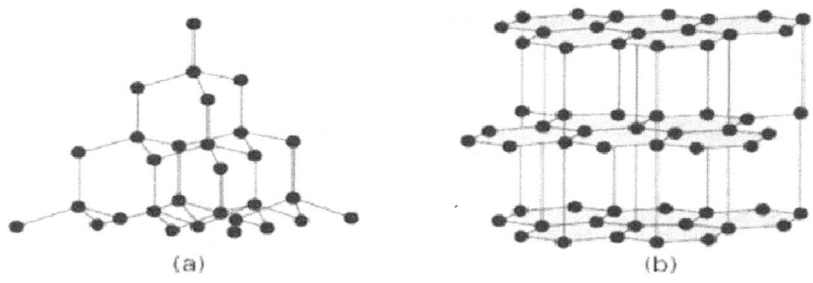

*Figure 4-3. The structure of diamond (a) and graphite (b)*

## 4.6. Boscovich's signposts to neutrino, gluons and quarks

"Also, in some of these classes (of particles), the absence of any force may be admitted; then the substance of one of these classes will pass perfectly freely through the substance of another without any collisions" /8, Section 518/. We only need to call these particles "neutrinos", and then Boscovich's idea is the same as contemporary understanding.

P. M. Rinard /26/ indicates that Boscovich's Theory can be linked to the modern theory of quarks, which is described by Dadich /2, p. 128-130/. Also Boscovich's Theory is important in relation to contemporary theory of elementary particles /6b/ and gluons /27/.

In 1993, Nobel Prize laureate Leon Lederman wrote that Boscovich "had an idea, completely crazy for the eighteenth century (and possibly any other)... Boscovich argues, no less, that matter is composed of the particles have no dimensions! We found a particle just a couple of decades ago that fits such a description. It's called a quark" /83, p. 103/.

# 5. CONFIRMATION OF BOSCOVICH'S FORCE LAW IN MODERN SCIENCE /23, 40/

## 5.1. Relation of force and energy dependence on the distance between the particles

By Boscovich, the elementary points, particles of first order, then of the second and third order, atoms, molecules, series of atoms... are only certain levels in the hierarchy of matter. According to him all worlds of smaller dimensions are like a single point in relation to the larger world. It is believed that for every pair of particles in any level of the hierarchy of matter applies some form of curves as shown in Figure 3-1.

In order to check if Boscovich's Theory is correct, the crucial question is: **Has modern science confirmed that the interaction between particles at different levels of the hierarchy of matter is really described by Boscovich's curve**? By overview of the many papers on Boscovich's life and works /1-7/ we did not find that anyone asked that question, or has tried to give the answer. (A few examples of correctness of Boscovich's curve are stated by Dadich /2/, but these examples he took from the author of this monograph.) If the results of modern science do not confirm the validity of Boscovich curve – then we can speak about Boscovich's Theory only as a transient phase in the history of science, though taking into account he is our countrymen. If modern science confirms that it is correct, then we need a different approach to this Theory (than treating it as merely transient).

Therefore, we examined the way in which modern science interprets the interaction of particles depending on their distance. While for Boscovich the interaction presents the change of attractive and repulsive **forces**, in the current literature that interaction usually is presented as a change of **potential energy** with the distance of particles. However, bearing in mind that the force (**F**) is actually a negative value of differential change in energy (**E**) with distance (**r**), i.e. **F = - dE/dr**, both curves are oscillating and are very similar (Figure 5-1), and can be derived from each other. The distinction is that the stable and unstable distances (limits of cohesion and non-cohesion, i.e. intersections with the abscissa on Boscovich curve, Figure 5-1, below) correspond to minima and maxima of potential energy (Figure 5-1, above), which are stable and unstable distances according to contemporary interpretation. Therefore, by observing the change of potential energy with distance between the particles, one can infer that the change in force, and thus check the validity of Boscovich's curve.

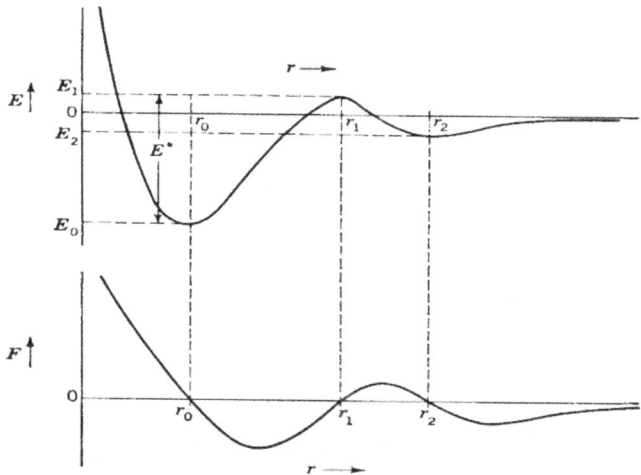

*Figure 5-1. The change of the potential energy (E) and force (F) depending on the distance (r) in the chemical reaction of the two atoms /28/*

## 5.2. Interaction of atoms

In Fig 5-1 it is actually displayed the change in the potential energy of a chemical reaction of two atoms. We see that the curve has a similar shape as Boscovich's curve (Figure 3-1c).

Interaction of atoms may not be chemical, but can be physical in nature. In this case, modern science has confirmed that this interaction is presented by a curve shown in Figure 5-2 such as in the case of sodium. In addition to liquid sodium, similar examples of argon and aluminium atoms are listed in Croxton's book /29/.

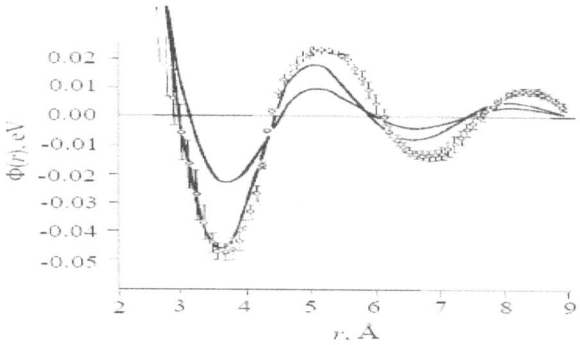

*Figure 5-2. The change in potential energy with the change in distance between atoms of liquid sodium /29/: Solid lines are theoretical curves, and the points are the experimental data*

Portnoy et al /30/ obtained similar results and showed the similar oscillatory curve for boron, magnesium, sodium, lead and aluminium. The fact that authors put that curve on their book cover (Figure 5-3) suggests to which extent the authors give significance to these findings.

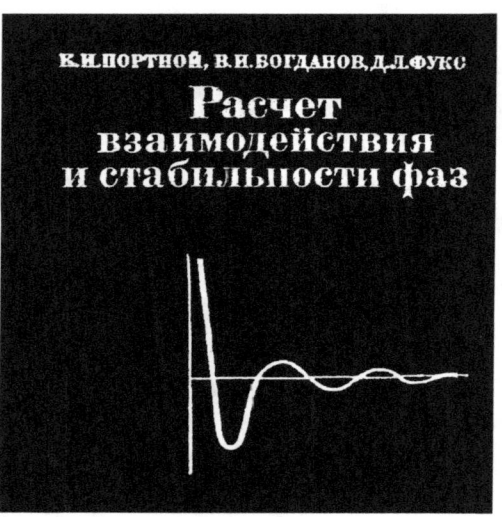

*Figure 5-3. Book cover of Portnoy et al /30/ (The curve represents the change in potential energy with the change in the distance between the atoms.)*

Croxton and Portnoy in their books do not call it Boscovich's Theory. Probably they are not familiar with it. However, it is obvious that these curves are generally identical with Boscovich's curve (Figure 3-1).

## 5.3. Interaction of molecules

Because molecules are formed by combining atoms, the molecules are at a higher level of the hierarchy of matter with respect to the atoms. According to Boscovich's Theory, the same law of force should be valid. Modern science has many examples that confirm this. Usually it is presented by a curve as in Figure 5-4, which fully corresponds to the curve of Boscovich in Fig 3-1b. Although this form of curve Boscovich showed since 1745, in his work "On the live forces" ("De viribus vivis"), today almost nobody mentions him. The exception is Kaplan, who indicates that Boscovich was the first who gave the law of interaction between particles /31/.

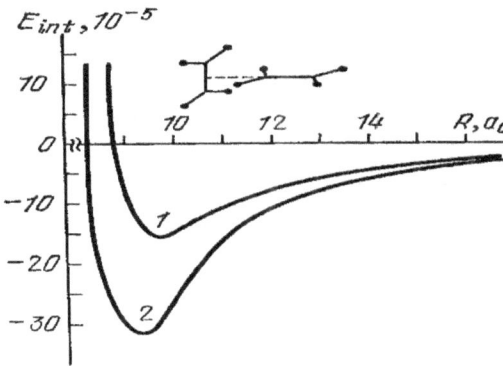

*Figure 5-4. The curve of the potential energy $E_{int}$ depending on the distance R between the two molecules of ethylene /31/. (Curves 1 and 2 are calculated theoretically by different methods. $E_{int}$ is in atomic units (AU), where AU = $2.6253 \times 10^6$ kJ/mol, the R has units $a_0$, where $a_0 = 5.2918 \times 10^{-7}$ cm.)*

In the adsorption of molecules on a solid surface, first occurs physical and then chemical adsorption. This change is attributed to the different interactions of the support's molecule with the adsorbing molecule, represented by a curve in Figure 5-5, which is the same as Boscovich's curve in Figure 3-1c.

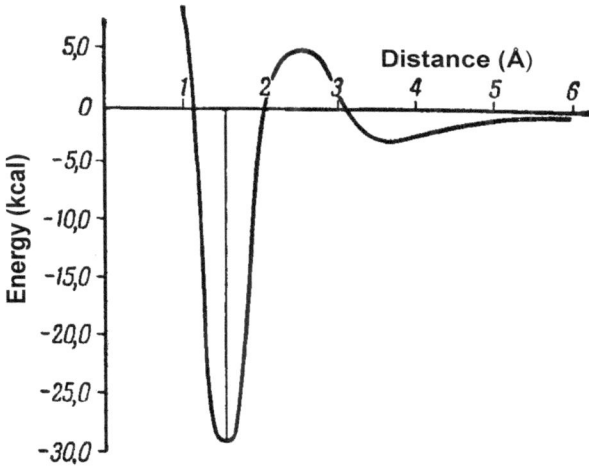

*Figure 5-5. The transition of the physical to the chemical adsorption /32/; at the abscissa is the distance of adsorbing from the surface of the carrier, and the ordinate is the change of potential energy; (1kcal = 4.184 kJ, 1Å = $10^{-8}$ cm)*

## 5.4. Interaction of nano-particles

Particles having a size of several tens to several hundreds of nano-metres are denoted as nano-particles. These are complex particles, formed by connecting together atoms or molecules. According to size and complexity nano-particles surpass that of the molecules; hence, they represent a higher level in the hierarchy of matter compared to the lower level of molecules. Pure nano-materials have great properties, and when added to other materials (atomic or molecular structure) significantly alter the properties of these materials, which has caused a great deal of attention in those working in the science and technology of materials.

Boscovich in his Theory pointed to the possible existence of such particles (section 4.5.). If someone could ask him how he would describe the interaction of these nano-particles, he would presumably say: "Well, by my curve!" Indeed, a recently published theoretical analysis and computer simulation /33/ showed that Boscovich was right (Figure 5-6).

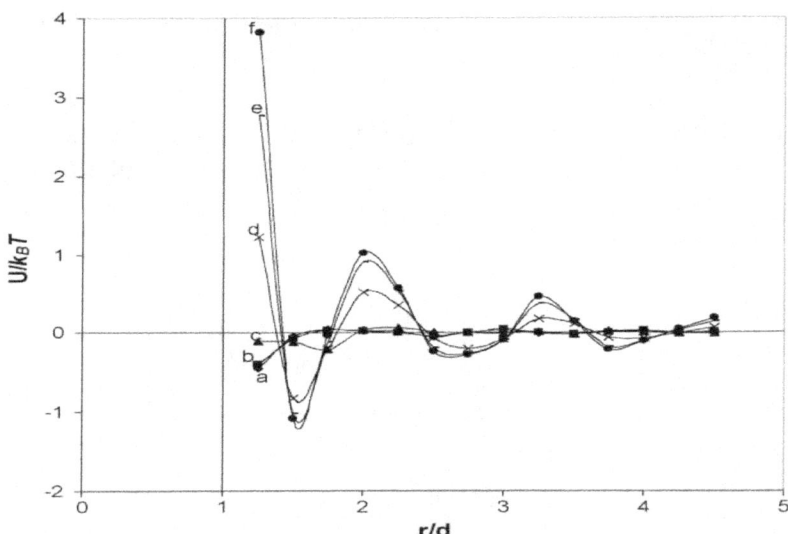

*Figure 5-6. Effective interaction potential energy (U) between identical charged nano-particles against the distance (r/d) between the centres of particles. (d is the particle diameter; individual curves refer to different values of charge.) /33/*

## 5.5. Interaction of macromolecules

A macromolecule is formed by chemical coupling of a large number of small molecules. The number of small molecules can be a few hundred to several millions. These also pose a particular level in the hierarchy of matter.

Boscovich in his Theory pointed to the possible existence of macromolecules (Chapter 4.4.). If someone could ask him to show what the interaction of macromolecules would look like, presumably he would again reply: "Well, by my curve!" Indeed, the results of modern science confirm /34/ and show that Boscovich would have been right (Figure 5-7).

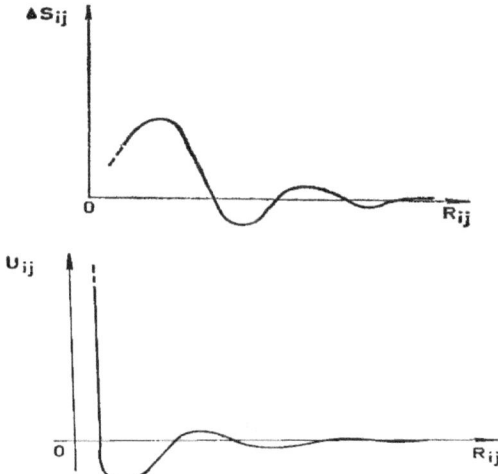

*Figure 5-7. The change in entropy ($S_{ij}$) and the enthalpy ($U_{ij}$) with distance ($R_{ij}$) between two macromolecular chains (i and j) /34/*

## 5.6. Interaction of colloidal particles

Interaction of colloidal particles (Figure 5-8) /35/ is also described by a form of Boscovich's curve (Figure 3-1c). The interaction of two clay particles can serve as an example (Figure 5-9) /36/.

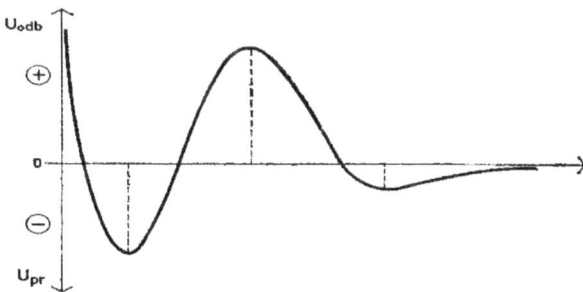

*Figure 5-8. Potential energy change with distance of two charged colloidal particles /35/*

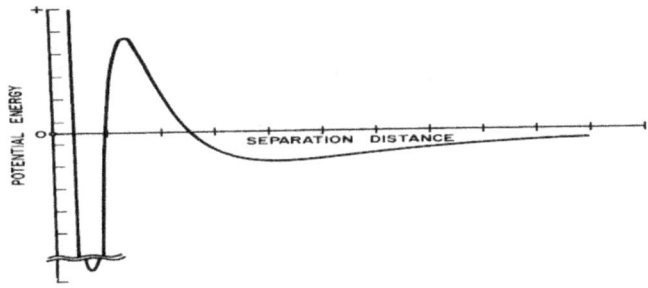

*Figure 5-9. Potential energy depending on the distance between the two particles of clay /36/*

## 5.7. Fission of heavy atomic nuclei

We have shown that in modern sciences the interactions of particles at different levels of the hierarchy of matter – from atoms, molecules, nano-particles, macromolecules and up to colloids – are described by Boscovich curves. In this and subsequent chapters we consider interactions of particles at the lower level of the atom.

The fission of heavy nuclei has the change of potential energy curve as an oscillatory shape (Figure 5-10) similar to Boscovich's curve. But, at high deformation of nuclei the potential energy has no horizontal asymptote as that which is shown in Boscovich's curve in Figure 3-1. However, Boscovich predicted such a possibility (/8/, Figure 14) as shown in Figure 5-11.

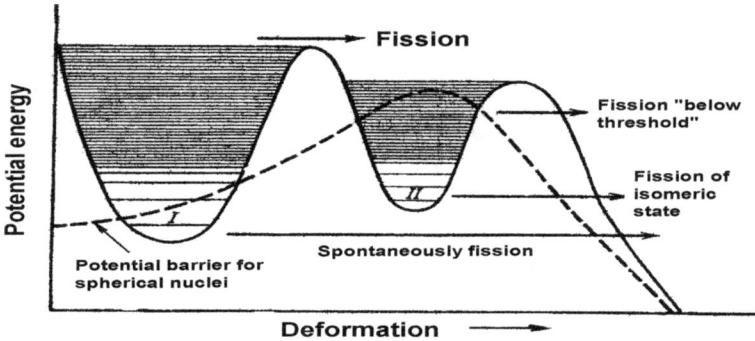

*Figure 5-10. Change of potential energy by fission of heavy nuclei of atoms /37/*

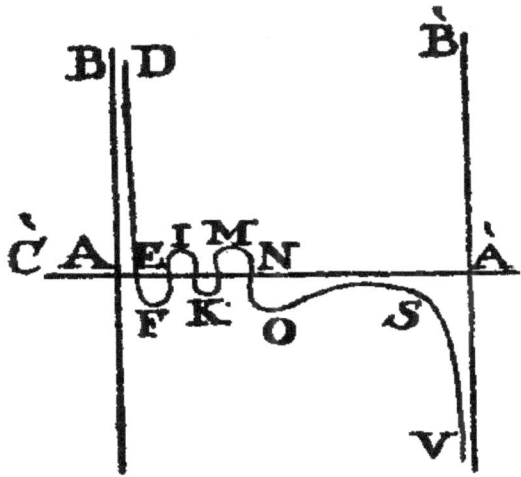

*Figure 5-11. One shape of Boscovich's curve similar to the curve in Figure 5-10.*

## 5.8. Energy of atomic nucleus

The dependence of atomic nucleus energy on its relative density in some cases may have an oscillatory shape (Figure 5-12) /38/. Taking into account that the distance between the nucleons is inversely proportional to density, it is clear that the presentation of nucleons interaction as an oscillatory shape is in line with Boscovich's curve.

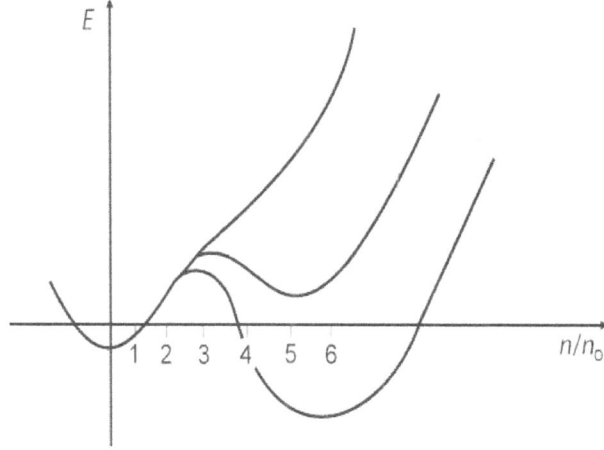

*Figure 5-12. The dependence of potential energy of atomic nucleus, on the ratio of actual density (n) to equilibrium density ($n_0$) /38/*

## 5.9. Interactions of nucleons and Λ° hyperon

Λ° hyperons are particles that belong to the baryons; they are uncharged with mass 2184 times that of the electron, spin ½, and lifetime $2.5 \cdot 10^{-10}$ seconds. By interaction of nucleons and Λ° hyperons there is an attractive force at large distances, but with a decrease in the distance it becomes a repulsive force and then attractive again, and at smaller distance it again becomes a repulsive force /39/. Curve for this interaction is not shown in /39/, but according to this description, it is exactly as described by Boscovich's curve (Figure 3-1c).

## 5.10. Conclusion concerning the validity of Boscovich's curve

In the previous sections we have discussed the way in which modern science describes interactions of particles at several levels in a hierarchy of matter. We have listed a few examples for each level (Table 5-1). In addition to the examples mentioned in this monograph, there are many others. In all cases, it has been shown to correct with Boscovich's curve.

Table 5-1. Outline of examples that confirm the validity of Boscovich curve

| Interactions in the hierarchy of matter | | |
|---|---|---|
| Level | Particles | Figures |
| 1 | Colloidal particles | 5-8. and 5.9. |
| 2 | Macromolecules | 5.7. |
| 3 | Nano-particles | 5.6. |
| 4 | Molecules | 5.4. and 5.5. |
| 5 | Atoms | 5.1., 5.2. and 5.3. |
| 6 | Nucleus and electrons | 4.1. |
| 7 | Fission of heavy nuclei | 5.10. |
| 8 | Λ° hyperon and nucleons | Description in Chapter 5.9. |
| 9 | Nucleons in nucleus | 5.12. |

Boscovich's Theory by its philosophical approach therefore made a great contribution to the development of modern science. By elaboration of his approach to the structure of matter and clarifying concepts in the hierarchy of matter, many natural laws have become very simple and generally applicable.

# 6. COMPRESSION OF MATTER – REFLECTIONS OF BOSCOVICH'S THEORY IN SAVICH-KASHANIN THEORY /41/

## 6.1. Introduction

There was a gap of two centuries since the creation of Boscovich's "Theory of natural philosophy" and the emergence of Savich-Kashanin theory /42-44/. The first Theory is based on Leibniz's monads, the law of continuity and the principles of classical Newtonian mechanics, improving and complementing them where Boscovich thinks it is necessary. The second of these theories has its basis in quantum mechanics, which it might seem very distant from Boscovich's Theory. And when taken into account that Boscovich applies his law of forces to the primary elements of matter (that are non-extended indivisible points), as well as to the particles of the first and second order (which are comprised of the primary elements), whereas Savich and Kashanin focussed their observations on particles "from the atom to the celestial bodies" /44/, then it might appear that the subjects of these two theories are completely different. We are left to wonder: is the Slavic origin of these authors the only link between the two theories?

No, it is not so. There is much common to both theories. First of all, there is the dialectical base of both theories, which we previously reported /45/. Here we only point out the similarities of these theories in interpreting compressing substance. To this purpose, we will just briefly outline some key postulates of both theories and compare them.

## 6.2. Material density changes according to Boscovich's opinion

The density of matter is the relationship between the mass and volume of a body. The definition applies in the case when there is a very large number of material particles, which occupy a certain space. Even then, Boscovich's force law has relevance; each set of two particles obeys his force law. If we consider the compression of dispersed points of matter, then, by Boscovich, **the density of the body gradually changes without any jumps** /8, Section 51/. Bouncy density change is not possible, because "if a given density persists for an hour, and then is changed in an instant of time into another twice as great, which will last for another hour; then in that instant of time which separate the two hours, there would have to be two densities at one and the same time, the simple and the double..." /8, section 52/. The body, that has two densities, is inconceivable. So, by Boscovich, there is no abrupt change in density.

To what extent can the compression of matter go? Boscovich argues that "as there is no limit to increase of rarity, so there are no limits to increase of density" /8, Section 89/.

## 6.3. Material density changes according to Savich-Kashanin theory

In contrast to Boscovich views, Savich and Kashanin believe that by the compressing of matter, it alternates between intervals of gradual and abrupt changes in density (Figure 6-1) /44/. The density of matter is gradually changed from $d_1^0$ to $d_1^*$ in the pressure range from $p_0^*$ to $p_1^*$. Then at the pressure $p_1^*$, there is a jump in the change of density from $d_1^*$ to $d_2^0$. Again, up to the pressure $p_2^*$ there is the interval of gradual change of density, and again there is a jump of density, etc... Substances can only have those values of density that correspond to intervals of 1, 2, 3, 4... Each interval corresponds to one phase state of matter. The density is gradually changing within a definite phase state. The transition from one phase to another is like a jump in terms of changes in density.

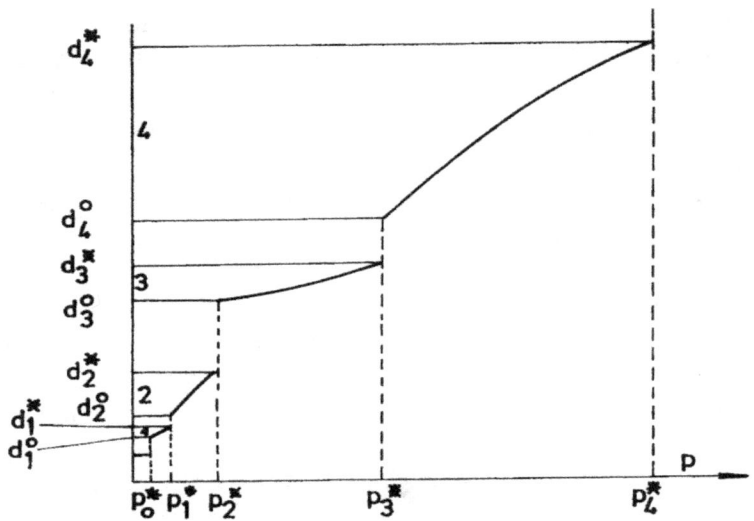

*Figure 6-1. Changes of density of matter (d) with the change of pressure (p) according to Savich-Kashanin theory /44/*

For the causes of these alternating stepwise and gradual changes in the density of matter, Savich and Kashanin looked to the combination of the type of quantum-mechanical phenomena, which describes the structure and properties of atoms. When atoms approach each other, there arises the moment when the atoms are close enough to each other that their outer electron orbit "touch" (Figure 6-2). Further compression is possible only if the electrons leave its former path and rebound from the atoms seeking a new space for their movement. Atoms, stripped due to these runaway electrons, can further approach each other until again "touch" the remaining outer electrons.

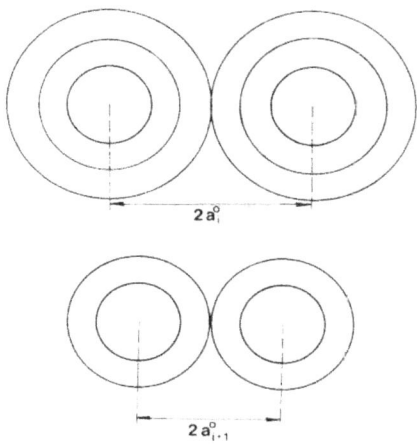

*Figure 6-2. Showing stepwise (i.e. act of leaping) atoms approaching each other under pressure where it is observed a stepwise change of the radius of action ($a_i^0 > a_{i+1}^0$) /44/*

"Excitation and ejection of electrons under the influence of pressure leads to a number of new phenomena in macro-systems. We see (Figure 6-2) that large and ultra-high pressure disrupts the inner micro-structure of the electron shells of chemical elements by pushing and ejecting electrons from them. Since the electrons are deployed by discrete, spaced levels, which are sharply separated from each other..., their ejection under this pressure will be in jumps. Accordingly, **material densities under pressure must be changed in jumps or sharp transitions from one value to another**. Due to the layered structure of the electron shells, by the displacement and ejection of electrons by the pressure, densities of the materials, as well as the properties of the macro-systems of particles, must exhibit abrupt changes" /44, p. 70/.

## 6.4. Relation between Boscovich's opinion and the opinion of Savich and Kashanin /41/

The density of a body is gradually changed without any jumps – claims Boscovich. Densities of the material under pressure must be change in jumps – claim Savich and Kashanin.

Boscovich, as 18th century citizen, built his comprehension of Nature on the law of continuity and classical Newtonian physics. Savich and Kashanin, our contemporaries, find support in modern quantum mechanics, which almost completely squeezes out the law of continuity and Newtonian mechanics from the micro-world. Therefore, Savich and Kashanin do not call on Boscovich's Theory, although they knew it /46/. Which of them is right – Boscovich or Savich and Kashanin?

Boscovich discovered his force law by reasoning, analyzing collisions between bodies (as well as other natural phenomena). However, he did not know the real cause why there was alternating turns in repulsive and attractive forces.

The above observations, however, indicate that there are missing links between the seemingly contradictory theories. The relation between Boscovich's force law and Bohr's model of the atom is accomplished through the works of Thomson, Rutherford and Bohr (Chapter 4.2.). This is the first link of Boscovich to Savich, as the Bohr model of the atom served for Savich as a display for the interpretation of abrupt changes of densities. The second link is calculation of the potential energy change when two atoms approach each other, a calculation based on a quantum-mechanical model of the atom. Next, the third link is the application of the results of this calculation for the prediction of properties of different phase states of matter made up by the atoms. These last two links are described in the books of Croxton /29/ and Portnoy /30/, although the authors do not cite papers of Boscovich, Savich and Kashanin. However, it can be seen that the change of potential energy as the distance changes between two atoms, with respect their discrete quantum-mechanical structure (the phenomena that Savich and Kashanin take into account), is described by the curve (Figure 5-2 and 5-3), which is identical with Boscovich's curve (Figure 3-1).

Thus, the path from Boscovich, through Thomson, Rutherford and Bohr up to Savich and Kashanin is connected completely. Moreover, Boscovich carefully analyzes his curve (Figure 3-1) and observes the intervals in which the particles spontaneously condense and the intervals in which the condensation is only possible if there is external pressure /8, Sections 190-194/. For example, if two atoms are at distance R (Figure 3-1), a further rapprochement of these atoms is possible only if it is exerted by external pressure, which is high enough to overcome the repulsive arch RQP. When the atoms come closer than P, there is attractive force acting on them. These atoms are still spontaneously and rapidly approaching and require no additional external pressure. When they come at distance less then N, again there appears a repulsive force, and again an outside pressure is necessary for the compression of matter. Boscovich's curve undoubtedly shows that the intervals of spontaneous and forced approach of particles are alternated.

There is an obvious similarity in the sense of Boscovich's curve (Figure 3-1) with Figure 6-1 of Savich and Kashanin. The ordinate of the first one (i.e. force) corresponds to the abscissa of the other (pressure, i.e. force over unit of surface area). The abscissa of the first one (i.e. distance) corresponds to the ordinate of the other (i.e. density, which is inversely proportional to distance). Furthermore, characteristic points on the Boscovich's curve (E, G, I, L, N, P, R), the so-called limits of cohesion and non-cohesion, are nothing else but the beginnings and endings of some steps in the diagram of Savich and Kashanin. Hence, **to every arch of the Boscovich's curve there is a corresponding step in the diagram of Savich and Kashanin.** A repulsive arch of Boscovich curve corresponds to gradual change (in density), but an attractive arch corresponds to a jump in density changes.

Then, does it not seem that Boscovich was wrong when he claimed that the change of density of matter must be a continuous, without jumping? Actually not! Nowadays it is known that the density of a substance significantly changes in the transition from solid to liquid or from liquid to a gaseous state (first-order phase transitions). But there are also known examples of phase changes of the second and higher order, in which the density is gradually changing when going from one phase state to another phase state, passing through all intervening values. But in both cases, the change of phase state is achieved as small fractions of the matter pass from one to another phase state. (The cloud does not condense entirely at once, but drop by drop.) On the micro-level the change is abrupt, like a jump. And that is what we describe by the Boscovich curve and diagram of Savich and Kashanin. But at the macroscopic level, the density change is gradual. Therefore, as to the question posed in this chapter, who is right, Boscovich **or** Savich and Kashanin, there is a dialectical answer: **both** Boscovich **and** Savich and Kashanin.

## 6.5. Density change in the compression of matter by the model of Savich and Kashanin

Savich and Kashanin in the form of a staircase diagram show the function, which describes the change in density of matter at the beginning and at the end of certain phases (Figure 6-3).

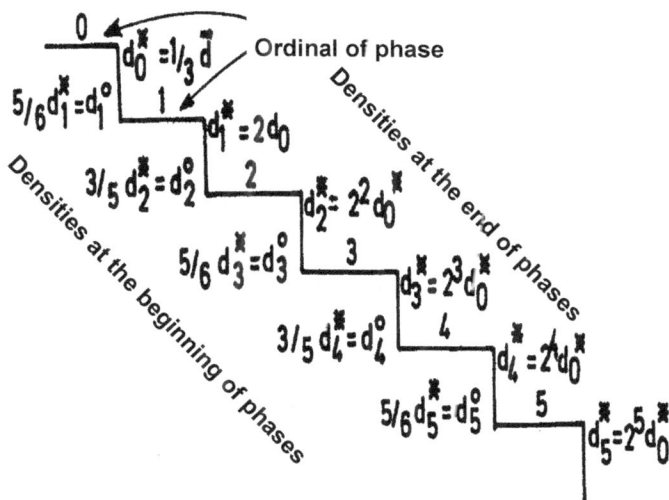

*Figure 6-3. The density at the beginning of $d_i^{\,\prime}$ at the end $d_i^*$ of individual phases (i = 1, 2, 3...) according to the Savich-Kashanin theory /44/*

Law of Savich and Kashanin for the stepwise change of density is not directly derived from the quantum-mechanical model of the atom, but it was only assumed that abrupt changes in the macro system are caused by the stepwise changes in Microsystems. It should be noted that Savich and Kashanin empirically come to the law that the density of matter (d) at the end of individual phases (i) changes abruptly, according to the expression (6-1a). Densities at the beginning of the individual phases (i) are calculated by dividing the density at the end of a certain phase with parameter $\alpha$, where $\alpha = 5/3$ and $\alpha = 6/5$ for the even and the odd phases, respectively (6-1b and 6-1c).

$$d_{i-1}^* = 2 d_i^* \tag{6-1a}$$
$$d_i^0 = d_i^*/\alpha \tag{6-1b}$$
$$\alpha = 5/3 \text{ and } \alpha = 6/5 \text{ for the even and the odd phases, respectively} \tag{6-1c}$$

Savich and Kashanin calculated these values of parameter $\alpha$ by taking into account the van der Waals equation of state for real gases (6-2).

$$(P + a/V^2)(V - b) = RT \tag{6-2}$$

P, V and T are the pressure, volume, and absolute temperature of gas, respectively, **a** and **b** are the so-called van der Waals constants, R is the universal gas constant.

It follows from equation (6-2) that the constant b, so-called "covolume", is the volume $V_0$, which a gas would have at absolute zero (T = 0 K). It is also known that it follows from this equation that covolume b is equal to one third of critical volume $V_c$, for given material at critical point (6-3). ($V_0$ and $V_c$ are characteristic points in P-V-T diagram, Figure 7-1a.) Respecting the van der Waals equation of state, Savich and Kashanin took into account the relationship (6-3) and calculated the values of parameter $\alpha$ as in equation (6-1c). It should be noted that this relationship (6-3) is one of the important assumptions built into obtaining the mathematical model of Savich and Kashanin (Figure 6-3).

$$b = V_0 = V_c/3 \tag{6-3}$$

It should be borne in mind that the specific volume of matter (V) is equal to the reciprocal of the density (d) (6-4), so it is easy to calculate the value of one of them if the other is known.

$$V = 1/d \tag{6-4}$$

Another important assumption of Savich and Kashanin used to derive their mathematical model is that matter at the end of the zero-phase has volume, or density, which corresponds to the critical point (6-5).

$$d_0^* = 1/V_c \tag{6-5}$$

## 6.6. Mean densities of planets in Solar system calculated by model of Savich-Kashanin

Savich and Kashanin applied the above mathematical model (Figure 6-3) to calculate the mean density of planets in the Solar system and the results of their calculations they compared with astronomical data available in that time. According to their calculations some planet should have a density of 0.67 g/cm$^3$, which approximately corresponds to the density of Saturn (0.65 g/cm$^3$). For one group of planets the calculated density was 1.33 g/cm$^3$; that corresponds to Jupiter (1.34 g/cm$^3$), Uranus (1.36 g/cm$^2$) and Neptune (1.32 g/cm$^3$). For the second group of planets the calculated density was 5.33 g/cm$^3$; that which corresponds to Earth (5.52 g/cm$^3$), Venus (5.21 g/cm$^3$) and Mercury (5.6 g/cm$^3$). The agreement of calculated and measured values is very good. A large discrepancy is only in the case of Mars: the calculated value is 5.33 g/cm$^3$, while the empirically estimated value is 3.94 g/cm$^3$. Savich and Kashanin believed that their calculation was correct, and the discrepancy with the observed value of the density indicated a possible error of astronomical data for the radius of Mars.

## 6.7. Adaptation of stepwise mathematical model by actual empirical data

Analyzing the mathematical model of Savich and Kashanin we noticed that some of their assumptions are not consistent with recent empirical data.

Empirical data show that the relations given by (6-3) are not correct, but that the volume of material at critical point ($V_c$) is twice the value of covolume (b) /47/, a four-fold higher value than the volume of matter at absolute zero temperature ($V_0$) /48/ (6-6).

$$V_c = 2\,b = 4\,V_0 \qquad (6\text{-}6)$$

Analysis of the compressed gaseous ethylene showed that the different phases are indeed formed /53/. However, the density of ethylene at the critical point corresponds to the end of first phase, but not to the end of zero phase, as it is proposed by and Savich Kashanin in relation (6-5).

In the theory of Savich and Kashanin the initial state of matter is the rarefied gas that is condensed into forming sun and planets. This means that the beginning of the zero phase should have a density which is close to zero. However, in their staircase model (Figure 6-3) the density at the beginning at the zero phase has some definite value higher then zero, i.e. $d_0^0 = (3/5)\, d_0^* \gg 0$.

Due to these empirical facts, it was necessary to adapt the mathematical model so that it will be consistent with these empirical facts. This adaptation and theoretical derivation of our model is presented in /49/ which the ratio of the characteristic volumes of matter to the critical volume is shown (Chapter 7). The same model can be represented by the density of the matter, which is the reciprocal of the volume (6-4). Based on the above, the theoretical mathematical model was obtained that shows the relationship of mean density of the planet to mean density of the Sun (see Figure 6-4).

According to our staircase model, the condensation starts with the gaseous matter where the density is close to zero and then the density increases in the phase transition from zero to the first phase, then to the second, then to the third phase and so on. The coefficients α that describe the ratio of density of the end and beginning of some phases (6-7) are not 5/3 or 6/5, as in Savich-Kashanin theory (6-1c), but in some stages have the values according to the formula (6-8).

$$\alpha(i) = d_i^* / d_i^0 \qquad (6\text{-}7)$$

$$\alpha(i) = 2^{-1/i} \qquad (6\text{-}8)$$

Where the number of the phase is $i = -1, 0, 1, 2, 3...$

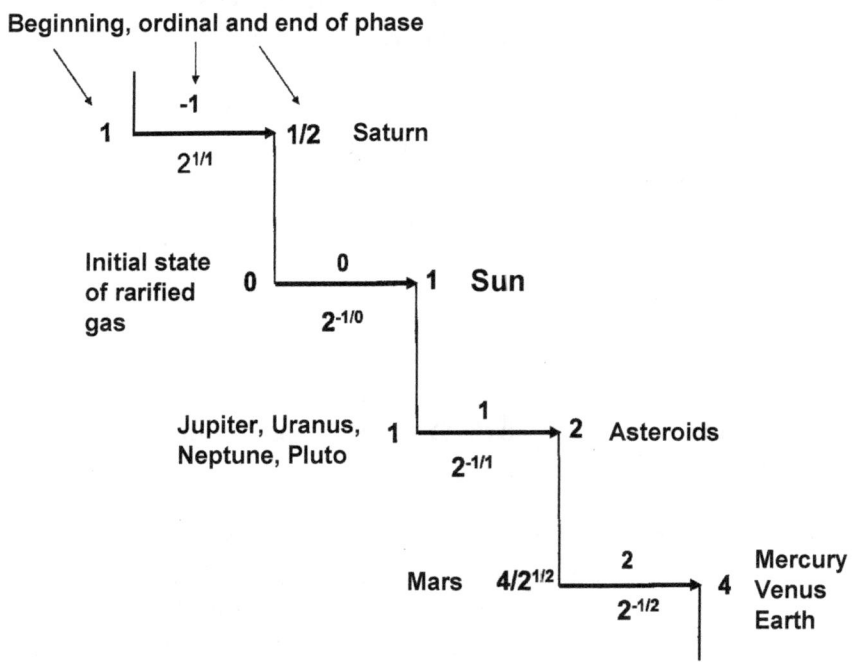

*Figure 6-4. Our theoretical staircase model – the ratio of mean density of the planets with that of the Sun ($d_s = 1.41$ g/cm³) /49/*

Bearing in mind that the mean density of the Sun is equal to 1.41 g/cm$^3$, the values of mean density of the planets can be calculated by our model (Figure 6-4) /50/. Agreement with empirical data /51/ was very good (Table 6-1). In addition, unlike the staircase model of Savich and Kashanin (Figure 6-3), our model shows that mean density of Mars should be about 4 g/cm$^3$, which is close to the empirical data.

Table 6-1. Mean density of planets in the Solar system

| Planet | Mean density (g/cm$^3$) | |
|---|---|---|
| | Empirical data /51/ | Calculated using our staircase model (Figure 6-4) /50/ |
| Mercury | 5.4 | 5.64 |
| Venus | 5.2 | 5.64 |
| Earth | 5.5 | 5.64 |
| Mars | 3.9 | 4.00 |
| Jupiter | 1.3 | 1.41 |
| Saturn | 0.7 | 0.71 |
| Uranus | 1.6 | 1.41 |
| Neptune | 1.7 | 1.41 |

Based on our staircase model, a celestial body (or bodies) with mean density 2.8 g/cm$^3$ could exist in the Solar system. Indeed, it corresponds to the asteroids whose density is in the range of 2.0 to 3.5 g/cm$^3$.

Although Pluto was recently removed from the list of planets, we also calculated by our model that its mean density would be 1.41 g/cm$^3$ which is in accordance with empirical data, i.e. 1.75 g/cm$^3$ (though in the literature one can find other very different values).

The trend of densification is indicated by an arrow in Figure 6-4. If Saturn followed the same trend, then there will be no condensation, but instead a state of spreading rarefied gas. In other words, according to our model, planet Saturn can not be condensed (i.e. is mostly a ball of gas).

There is a hypothesis that the sun has a twin star, which is called Nemesis. Mean density of Nemesis would be 79.21 g/cm$^3$ /52/. It can be calculated by extrapolation of our staircase model that there could be a body with density of 80.63 g/cm$^3$, which is in good agreement with the above data for Nemesis.

# 7. APPLICABILITY OF BOSCOVICH'S THEORY

## 7.1. Introduction

Boscovich pointed to numerous diverse application of his Theory in mechanics and physics. He considered the following phenomena: pressure in fluids, the speed of the fluid flow out of a container; equality of action and reaction; gravitation; cohesion; solid and fluid state; non-flexible, flexible, elastic and brittle rods; viscosity; fluid resistance; elasticity and softness; ductility and malleability; chemical operations; nature of fire; light and its properties; tastes and smells; sound; cold and heat, electricity and magnetism. These considerations were supported by scientific knowledge that existed in the 18th century, but which in many cases today is surpassed. So, some of Boscovich's considerations are therefore outdated and surpassed. However, by evaluation of his interpretations, always we should be cautious and ask ourselves: "Maybe Boscovich was right?"

Very important is the question: "Is Boscovich's Theory applicable and useful in modern science?" This question is quite reasonable when one takes into account that modern science has only recently confirmed the validity of Boscovich's force law (Chapter 5). Maybe Boscovich's Theory can not provide a solution of some scientific problems, but it can be a good clue and signpost to finding solutions by modern scientific methods.

The author of this monograph has several times used Boscovich as road signs to successfully find the answers to some open scientific questions. In this chapter there will be only a short summary, and a more detailed explanation is given in the cited published papers.

## 7.2. Meaning of critical volume of matter /49, 53/

Boscovich suggested the possibility that "it may happen that two points approaching one another from a long way off, but not exactly in the straight line joining them... then the points will not reverse their motion and recede, but will gyrate about a motionless middle point of space for ever more, always remaining very near to one another, the distance between them not being appreciable by the senses". /8, Section 201/. In this case, the repelling centrifugal force equals to the attractive centripetal force, and the distance of these rotating pair of points should correspond to the outermost cohesion limit (position R in Figure 3-1a), The two particles by their gyration will occupy a sphere, wherein the diameter of the sphere corresponds to a cohesion limit.

It is known that molecules in the gaseous state of matter have such large initial distance. If they approached each other they can form a rotating molecular pair at a distance that corresponds to the limit of cohesion (by Boscovich), i.e. corresponds to the minimum potential energy (according to the modern terminology).

The volume of the sphere in which the pair rotates can be calculated using a simple equation (7-1):

$$V_p = (2/3) \pi r_e^3 N \qquad (7\text{-}1)$$

where $V_p$ is the volume of sphere occupied by the rotating molecular pair, calculated on one mole; $r_e$ is equilibrium distance between the molecules; $\pi = 3.14$; N is the Avogadro's number ($6.022 \times 10^{23}$ molecules per mole).

If almost all the gas molecules were coupled as rotating molecular pairs at a distance $r_e$ corresponding to the outermost cohesion limit (point R in Figure 3-1a), then this would be a very characteristic state of matter in the pressure-temperature-volume thermodynamic diagram, presented in Figure 7-1a. One characteristic state is critical point above which it is not possible to condense vapor to liquid.

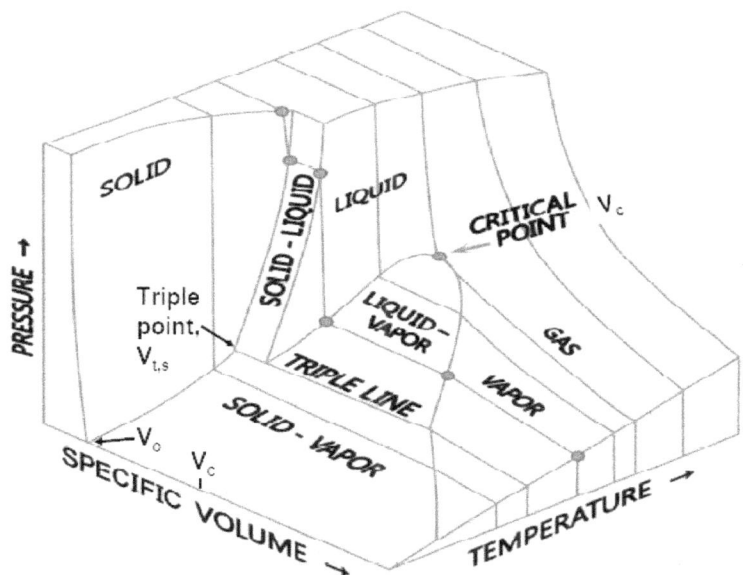

*Figure 7-1a. Characteristic points of matters, i.e. critical point, triple point and absolute zero temperature with corresponding volumes ($V_c$, $V_{t,s}$ and $V_O$), presented in a typical pressure-volume-temperature diagram of substances.*

There is empirical data for equilibrium distance $r_e$ for many substances and it is possible to calculate $V_p$ by equation (7-1). Also, there is empirical data for critical volume $V_c$ for many substances. We have proved for 92 substances that the volumes of rotating molecular pairs $V_p$ is equal to $V_c$ (Figure 7-1b). It means that the volumes of rotating molecular pairs at outermost cohesion limit (position R in Figure 3-1a) corresponds to their volumes $V_c$ at critical point.

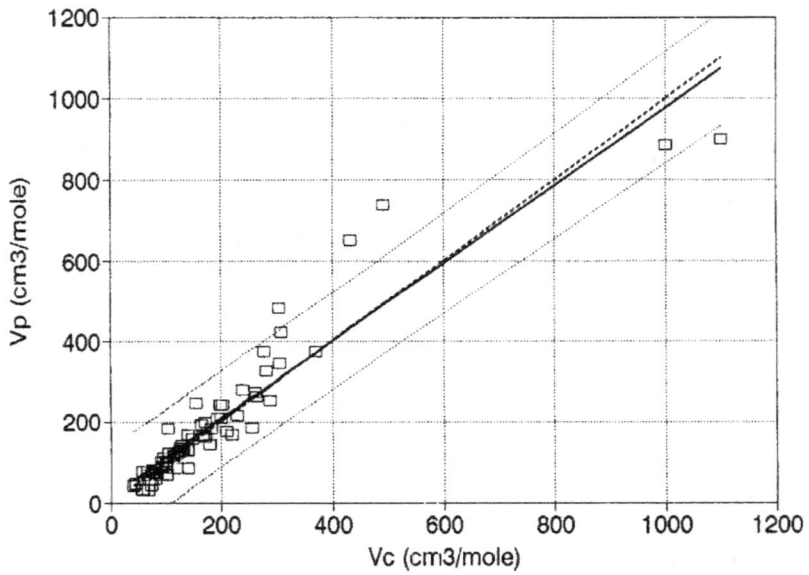

*Figure 7-1b. Equality of critical volume $V_c$ and the volume $V_p$ occupied by two rotating molecules at outermost cohesion limits (R in fig. 3-1a) /49, 53/. (□ empirical data for 92 substances; — Regression; ---- Expected; ····· Confidence limits. Standard error of 60.1 $cm^3/mol$; Correlation coefficient 0.94)*

## 7.3. Characteristic volumes of matter /49, 54/

It can be concluded that the outermost limits of cohesion has a certain physical meaning and corresponds to a certain distinctive state of matter, i.e. the critical point. It is reasonable to ask whether other limits of cohesion and noncohesion on the Boscovich's curve have some physical meaning. It was explained (Chapter 6) that each step in the diagram of Savich and Kashanin (Figure 6-3) corresponds to a repulsive and an attractive arch of the Boscovich curve (Figure 3-1). Based on the empirically established relationship (6-6) it can be concluded that the specific volumes of matter at the end of first, second and third phases correspond to the critical volume, covolume (van der Waals's constants b) and volume of matter at absolute zero. Once the adjustments of the coefficients α(i) are performed (equations 6-8), the volumes of the end of phases are multiplied by α(i) to obtain the volumes at the beginning of the phases (Figure 7-2).

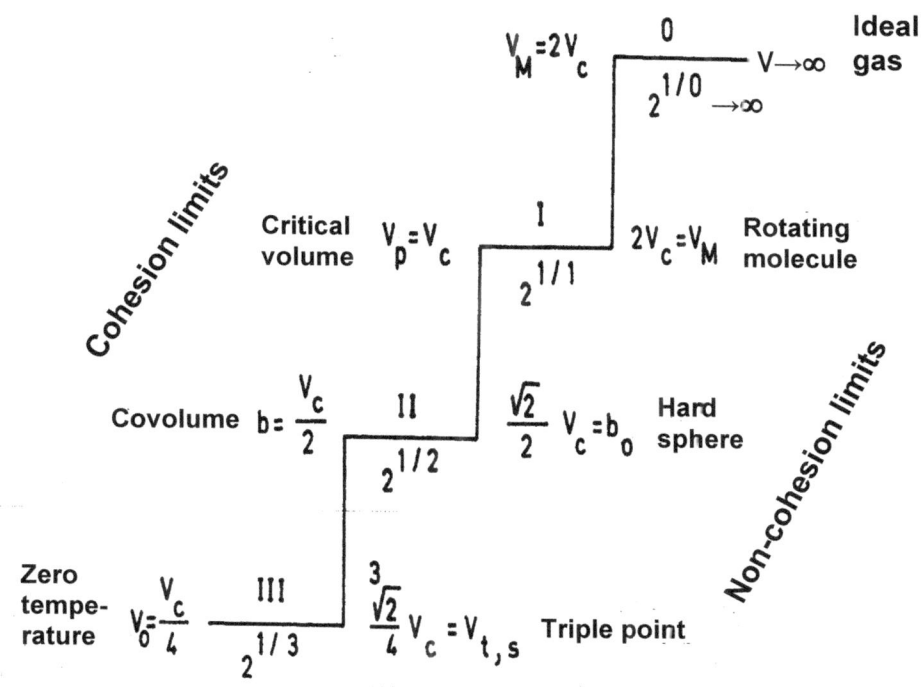

*Figure 7-2. The relationship between the specific volumes of matter in characteristic states and the critical volume $V_c$, i.e. our staircase mathematical model /49, 54/*

It has been shown that the volumes at the beginnings of a certain phases have some definitive physical meaning: $V_M$ is volume occupied by the rotation of individual molecules; $b_0$ is the hard sphere volume, occupied by the two molecules for the distance at which the potential energy is equal to zero; $V_{t,s}$ is the volume of the solid phase at the triple point. Agreement of calculated values with experimental value for specified volume is very good (Figure 7-3).

Although the gravitational and intermolecular forces are different nature of forces, the mathematical models presented in staircase diagrams in Figures 6-4 and 7-2, are completely identical. (They differ only in that the first diagram presents the calculation of density, while the second diagram presents the calculation of specific volume, i.e. the reciprocal value of density.) This identity of mathematical models confirms Boscovich's opinion that there is a one unique law of forces that exist in nature (Figure 3-1).

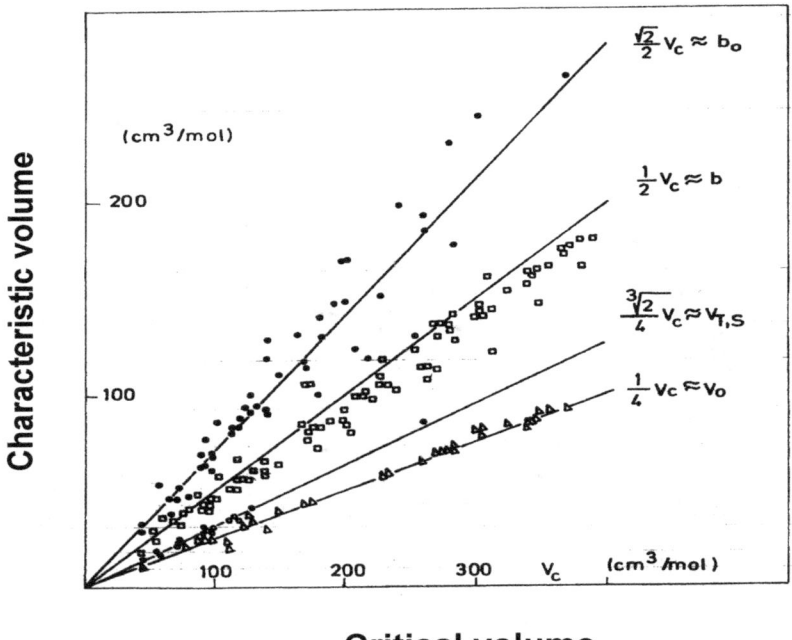

*Figure 7-3. The relationship between the critical volume $V_c$ and other characteristic volumes: hard sphere volume $b_0$, covolume b, the volume of the solid phase in the triple point $V_{t,s}$, volume of matter at absolute zero $V_0$. Lines represent the theoretical expected values based on our model (Figure 7-2), and points are experimentally determined values taken from the literature for the 143 substances: metals, inert gases, elements, saturated and unsaturated hydrocarbons, aromatic hydrocarbons, organic or inorganic compounds of oxygen, nitrogen, sulphur and halogens /49, 54/.*

The law of nature, according to which the density of matter at the end of certain phases is related to the expression $d_{i+1}^* = 2\, d_i^*$ (6-1) was spotted by Savich and Kashanin when analyzing the density of the planets in the Solar system; it has a much deeper meaning than mentioned by these authors. Interpreting the abrupt changes in the density of matter by step-transition of electrons from one orbit to another, Savich and Kashanin, unknowingly, in their theory built in Boscovich's law of interaction of particles of matter. Each step in the diagram of Savich and Kashanin corresponds to an arch of the Boscovich curve. This similarity and common dialectical-materialistic core of both theories, give to them such a type of universality, which is attributive only to those general laws of nature, on which foundations rests the magnificent edifice of modern science. Boscovich's Theory evolved from the interaction of "non-extended material points" leading to the quantum model of the atom, and this is carried on by the Savich-Kashanin theory, leading to describing how these atoms make up the celestial body and Solar system.

Hence, it is not surprising that our attempt was fruitful in incorporating the molecular and supra-molecular particles into Savich-Kashanin's steps and Boscovich's arches. So, we get the volumes at the beginning and end of each phase representing the characteristic volumes of matter. These are universal states of matter, unequivocally determined by the nature of matter itself and by Boscovich's unique law of forces existing in nature.

## 7.4. Physico-chemical state and polymerization of compressed ethylene gas /53, 55/

Ethylene molecule has a double bond and can be polymerized by a free-radical mechanism: the initial radical R• binds to a molecule of ethylene $CH_2=CH_2$ by breaking the double bond and a new radical is formed on just combined ethylene molecule (7-2). Then the next molecule of ethylene is bonded, and again the next, and so a few hundred or thousand times, resulting in a macromolecular chain of bonded molecules of ethylene. The resulting plastic mass, composed of such macromolecules, is referred to as polyethylene:

$$R• + CH_2=CH_2 \rightarrow R-CH_2-CH_2• \qquad (7-2)$$

The process was discovered in 1933, by Imperial Chemical Industry Company. The peculiarity of this simple polymerization is that it can be done only if the gaseous ethylene is compressed to very high pressure. Typical polymerization conditions in industrial plants are in the range 1000-3000 bar and 150-300 °C. Since its discovery, in the following decades the unresolved question was - why is it necessary to have such high pressure. Hunter /56/ observed that the density of compressed ethylene gas at the polymerization conditions is approximately 0.46 g/cm$^3$. This value exceeds the density of the randomly packed ethylene molecules which is 0.28 g/cm$^3$. The calculated average distance between the molecules of ethylene under polymerization conditions is 0.4-0.5 nm, which is less than the diameter of the molecule (0.5 nm). Hunter concluded that in these conditions the ethylene molecule is regularly packed, suitably oriented and distorted. He concluded that the compression achieves certain molecular organization of ethylene, which is a prerequisite for successful polymerization. However, he did not explain how the molecules are packed, oriented and distorted.

The interaction of most molecules, including ethylene molecules, is usually presented by Lennard-Jones's potential (Figure 5-4) published in 1924, which is similar to Boscovich's curve (Figure 3-1b) published in 1745. The empirical value for the distance between the centers of two molecules of ethylene at a minimum potential energy is $r_e = 0.466$ nm. Thus, **rotating molecular pairs of ethylene are formed.** The volume occupied by rotation of a pair is 127.6 cm$^3$/mol calculated by equation (7-1).

At shorter distances a strong repulsive force is expected, according to Figure 5-4. Ethylene, with its molecules at that distance, has density 0.22 g/cm$^3$. This value is two times lower than the density of ethylene in the polymerization conditions. This means that the molecules of ethylene may come towards each other at a distance less than 0.466 nm.

Therefore, we assumed /53, 55/ that instead of Lennard-Jones's potential (Figure 5-4), it is more appropriate (and hence correct) to apply Boscovich's curve as shown in Figure 3-1a. We supposed that the distance between the two paired molecules corresponds to position R in Figure 3-1a. Once the free volume between the molecular pairs is exhausted by compression, the additional compression is possible on account of space that exists between the two molecules in a pair. These already paired molecules come more closely towards each other, thus forming a **rotating bimolecule**. Van der Waals's constant b corresponds to the volume occupied by two molecules that have touched. Hence, constant b is named "covolume". Therefore, we proposed that a rotating bimolecule occupies volume b=57.1 cm$^3$/mol, which is the empirically determined value of ethylene covolume.

By additional compression, more and more molecules are transformed into bimolecules. The density of ethylene composed of bimolecules is $d_b$=M/b=0.49 g/cm$^3$, where M=28 g/mol is molar mass of ethylene. This density corresponds to that in ethylene polymerization conditions, mentioned above. Consequently, the formation of bimolecules is a clue to the prerequisite of successful polymerization.

Bimolecules rotate about all three axes, i.e. they have three degrees of freedom (3D). A further condensation is possible by cooling, which reduces the degree of freedom to one (1D). As a consequence, bimolecules are combined and **linear oligomolecules are formed**, which rotate around the longitudinal axis. As the dimensions of ethylene molecule and distance between two of these molecules is known, it is easy to calculate that the molar volume occupied by rotation of oligomolecule is 37.8 cm$^3$/mol.

On the basis of these assumptions the supra-molecular organization of compressed ethylene was proposed (Figure 7-4).

The existence of these particles and phase transitions of the second and third orders in compressed ethylene have been confirmed by thermodynamic, physical and spectroscopic methods /53, 55/. The phase transition from α to β phase occurs under conditions when the volume of ethylene is equal to critical ($V/V_c$=1), transition from β to γ phase is when entropy of ethylene is equal to the critical ethylene entropy ($S/S_c$=1).

We have shown that the polymerization of ethylene is only possible in the β and γ phase, and it has the highest rate at phase transition β-γ, when ethylene from less ordered β phase transforms to the more ordered quasi-crystalline γ phase.

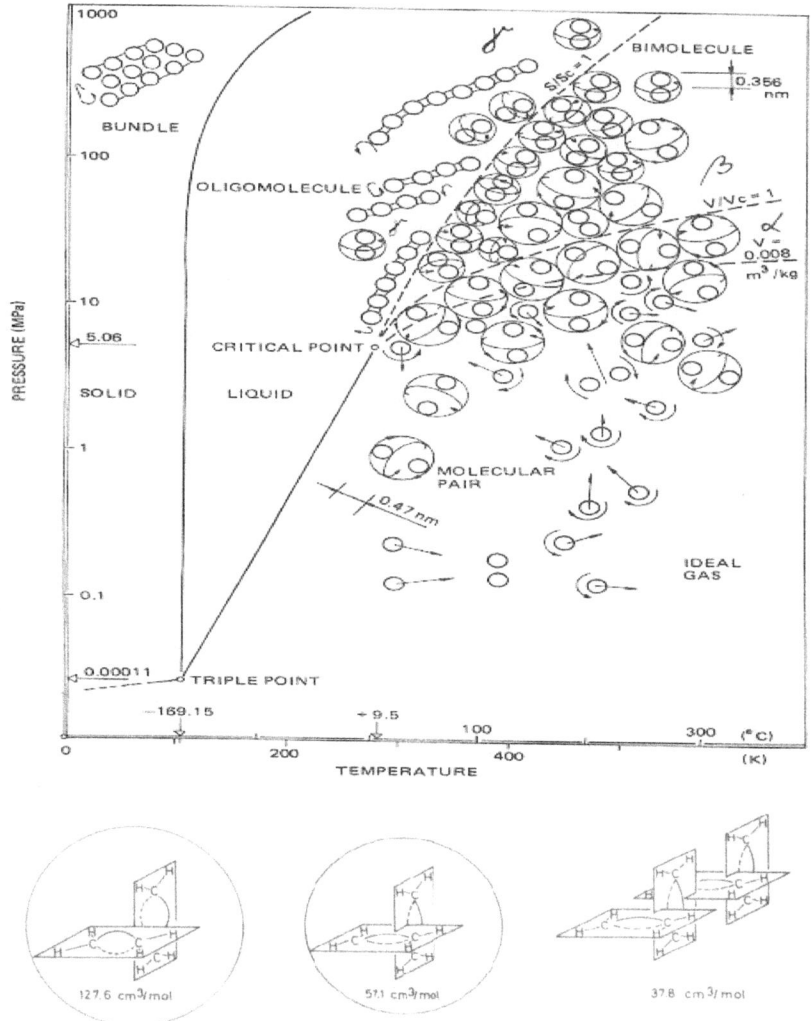

*Figure 7-4. The phase state of molecular particles and their volumes (empirical values) of compressed ethylene /53, 55, 57-60, 62-65/*

The ethylene supra-molecular particles have a decisive influence on the mechanism and kinetics of polymerization, as well as the structure and properties of the polyethylene. Crystalline polyethylene with regularly packed macromolecules originates in the ordered γ phase, while less crystalline polyethylene with disordered macromolecules originates in disordered β phase.

This interpretation, that has emerged thanks to Boscovich's roadmap, and its application in explanation of the polymerization of ethylene have been published in dozens of scientific papers /57-65/ and is of great importance for a better understanding and managing of the industrial polymerization process.

## 7.5. Effect of pressure on polyethylene melting point

The product of the industrial polymerization of ethylene by free-radical mechanism is a plastic material called low density polyethylene, which is in the form of solid partially crystalline granules. These granules are heated and melted in machines for making different products: electrical cable insulation, films and sheets, bottles, canisters, drums... At atmospheric pressure polyethylene melts at 115-120 °C. However, in the machine that melts polyethylene it is subjected to pressures from several hundred to several thousand bars. There is a danger that the melted substance solidifies at the increased pressure and causes damage to the machine or gives a defective product. Therefore, the melted substance should be warmed up well above 120 °C. Once, we (in Chemical industry "Panchevo") urgently required a proposal for the temperature at which to warm the melted substance, so that it did not harden at elevated pressures. At that moment in time, we did not have empirical data on the effect of pressure on the melting temperature of polyethylene.

However, in response to the above request, we found the signpost in Boscovich's Theory, i.e. his interpretation of the law of continuity /10, 11/. The law was first expressed by Leibniz, and Boscovich elaborated it in more detail. One of the consequence of this law is that "whenever the two variable quantities, which of course can change magnitude, are interconnected, then by the magnitude of one it can be determined the magnitude of the other" /10, section 102/.

We know that two such related quantities are: the degree of order of compressed ethylene (numerically presented by the entropy) and the degree of crystallinity of polyethylene, which is experimentally determined.

Furthermore Boscovich continues: "...let's imagine the two magnitudes of the first" (e. g. $E_i$ and $E_m$ in Figure 7-5) "and two magnitudes of the second quantity..." (e. g. $P_i$ and $P_m$) and "if the first quantity by a constant change passes from the first magnitude to the other, passing through all the [possible values of] magnitudes..." ($E_j...E_k...E_l$), then "it will happen also with the other quantity", i.e. it will pass through the proper magnitudes ($P_j...P_k...P_l$). If by $E_i$ we denote the γ phase, and by $E_m$ the β phase of ethylene, then $P_i$ and $P_m$ represents polyethylene, which is obtained by polymerization of the corresponding phases of ethylene. If $E_k$ is a transition from the γ to β phase of pure ethylene, i.e. "melting" of quasi-crystalline γ phase at a temperature which depends on the pressure, then $P_k$ must represent the corresponding phase transition, i.e. melting of polyethylene crystalline domains, which must occur at the same temperature and pressure. In other words, quasi-crystalline γ ethylene and partially crystalline polyethylene should melt at the same temperature and the same pressure.

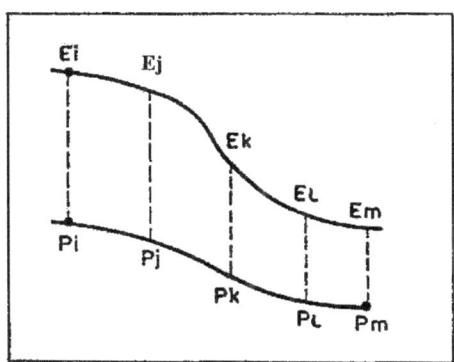

*Figure 7-5. Explanation of the application of the law of continuity on example of ethylene(E) and polyethylene (P) /53/*

We knew that γ ethylene at elevated pressures "melts" at a temperature at which the entropy is equal to the critical ($S/S_c=1$), the exact values of the temperature are known from thermodynamic data. We also concluded that the polyethylene should melt at the same temperatures as ethylene. The experimental data, collected afterwards, fully confirmed this conclusion (Figure 7-6) /25, 53, 62/.

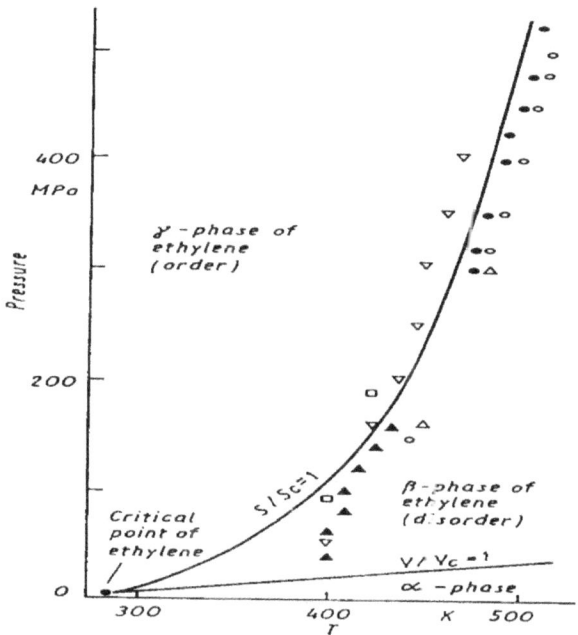

*Figure 7-6. Phase transition conditions β-γ in pure ethylene(line $S/S_c=1$) correspond to that of melting point of polyethylene (points) at elevated pressure /25, 53, 62/*

## 7.6. Structure of fluids based on Boscovich's Theory and Savich-Kashanin's theory

A huge problem in modern physics is a consequence of the fact that it can not accurately describe the structure of fluids, i.e. liquids and real compressed gases, in which plenty of very important physical, chemical, biological and other processes are performed. It is generally accepted that fluids have an amorphous structure; molecules in them are irregularly arranged. However, some short range local order is experimentally confirmed, i.e. some molecules are regularly arranged as consequence of different kinds of intermolecular forces acting between them. According to the comprehension of Eyring and Marchi, a liquid consists of two phases, a "gas-like" and the other phase that of "crystal like" domains /68/, but there is no explanation what is the fraction (percentage) of these domains, and how molecules are arranged in them.

Let's look at how Boscovich interpreted the interaction of particles in fluids. According to him /8, 71/ it can be described by the curve similar to Figure 3-1a, but instead of the outmost attraction arch, which represents gravitational force, there should be a repulsive arch since the gases have tendency to spread, i.e. there are repulsive forces between the molecules in gases. According to that description, S. Paushek-Bazhdar drew a solid oscillating curve (Figure 7-7) /84, 85/.

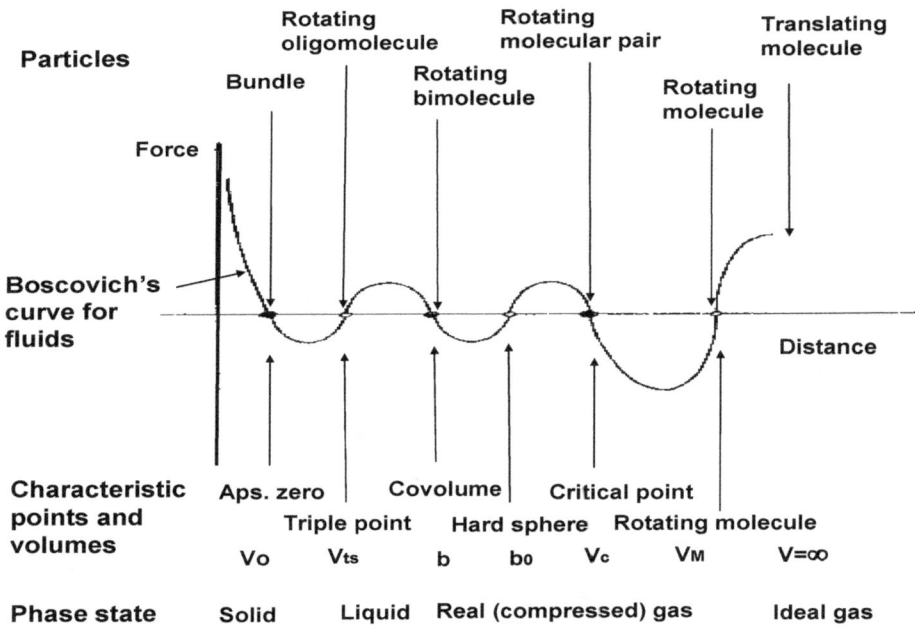

*Figure 7-7. Boscovich's curve for fluids /84, 85/ completed with supra-molecular particles, characteristic volumes of matter and phase states*

Previously, we explained that for every arch of the Boscovich's curve there is a corresponding step in the diagram of Savich and Kashanin (Chapter 6.4.). It was concluded that the molecules situated at cohesion and noncohesion limits on Boscovich's curve contribute to some characteristic state of matter that can be calculated by our staircase model (Figures 7-2 and 7-3). It was confirmed for 92 substances that the outermost cohesion limit corresponds to the critical point (Chapter 7-2). The anterior (i.e. earlier than the critical point) cohesion limit corresponds to the van der Waals constant b, i.e. covolume. Consequently, in Figure 7-7, we denoted the characteristic volumes of matter and the corresponding cohesion and noncohesion limits.

It was shown for 143 different substances that the same mathematical model can be used to describe the changes in characteristic volumes during condensation from ideal gas state (characterized by individual molecules having the highest mobility and freedom) to absolute zero temperature (where molecules are interconnected and totally immobilized) (Figures 7-2 and 7-3). If the same mathematical model can be used to describe these changes, the next two consequences can be logically drawn /49, 54/:

(A) **All substances are exposed to the same structural changes by passing from one characteristic state to another.** Otherwise, if the structural changes are not the same, the changes of volume for different substances could not be described by a common mathematical model.

(B) **All substances have the same supra-molecular structure in the same characteristic states.** This second conclusion is a logical consequence of the previous conclusion. If all substances are exposed to the same structural changes as they go from one characteristic state to another, it is quite reasonable to suppose that all substances have the same supra-molecular structure in the initial characteristic point, and some other structure in the subsequent characteristic point.

Next logical question is: What are these supra-molecular structures? The answer can be found in the case of compressed ethylene, and then can be generalized to other substances.

Based on our mathematical model (Figure 7-2), critical volume, covolume and volume of solid phase at triple point of ethylene were calculated: 127.6, 63.8 and 40 cm$^3$/mol, respectively. These values are very close to the empirically determined values of the volumes necessary for rotation of supra-molecular particles of ethylene, i.e. molecular pairs, bimolecules and oligomolecules: 127.6, 57.1 and 37.8 cm$^3$/mol, respectively. Hence, the supra-molecular structure of ethylene at critical point corresponds to molecular pair, since their volumes are equal, i.e. $V_c=V_p=127.6$ cm$^3$/mol. According to consequences (A) and (B), the same should be valid for other substances, i.e. supra-molecular structure of all substances at critical point should correspond to molecular pairs, and both volumes should be equal i.e. $V_c=V_p$. Indeed, it was confirmed that $V_c=V_p$ for 92 substances (Figure 7-1).

Also, the consequences (A) and (B) are valid for the beginning of phase 0 in staircase model (Figure 7-2), i.e. the structure of ideal gas, which is equal for all substances: chaotic movement of individuals molecules; average distance between molecules is the same for all substances, at same pressure-temperature-volume conditions (Avogadro-Ampere's law); all substances have the same energy equal P•V and all obey the unique equation of state, i.e. P•V=RT, where R is universal gas constant common for all gases.

The phase state and supra-molecular particles of compressed ethylene gas are presented in Figure 7-4, but only in supra-critical conditions, i.e. above critical temperature and above critical pressure. But, what is structure of liquid ethylene? It can be imagined by cooling and depressurizing of γ phase of ethylene. Consequently, the liquid ethylene should be a dynamic equilibrium of bimolecules and oligomolecules, the former rotating around three axes and the second around one axis, the former dominating at high temperatures, the latter prevailing at low temperatures. Bimolecules vanish at melting point. Indeed, Rytter and Gruen /86/ investigated the ethylene transition gas→liquid→solid by infrared spectra and interpreted data in terms of monomer→dimer→aggregate→crystal scheme.

According to consequences (A) and (B), liquids of other substances should have the same structure as liquid ethylene.

Indeed, Korolev et al /70/ have confirmed for 8000 organic compounds that their liquids consist of ordered and disordered domains: by cooling gas at the temperature of condensation the individual molecules pass into liquid as dimers. By further cooling linear oligomers are formed, and at even lower temperatures the molecules are all mutually connected forming a continuous associative structure.

Also according to Eyring and Marchi, liquid consists of two phases, a "gas-like" phase and a "crystal like" phase /68/. According to our conceptions, the "gas-like" domains consist of 3D-rotating bimolecules. The "crystal-like" domains consist of 1D-rotating linear oligomolecules. There is a dynamic equilibrium between these phases, i.e. bimolecules are transformed into oligomolecules and vice versa.

It should be noted, however, that Lennard-Jones's potential (Figure 5-4), which is usually used to represent the interaction of molecules in liquid, is not appropriate: two molecules at equilibrium distance $r_e$ occupy the volume equals to critical volume $V_c$ (Figure 7-1). The minimum volume that can be achieved according to this potential is $b_o$, i.e. so called 'hard sphere volume' which is equal to 0.71 $V_c$ (Figures 7-2 and 7-3). However, the actual volumes of liquids are much smaller, i.e. 0.3-0.5 $V_c$. Hence, the rotating molecular pairs at distance $r_e$, according to Lennard-Jones's potential, cannot represent dimers in liquid, that were found by Korolev et al. Molecules in dimers have to be at shorter distance where the volume is equal to van der Waals's constant b, i.e. covolume.

Consequently, the positions of molecular and supra-molecular particles in Boscovich's curve for fluids are as presented in Figure 7-7.

We applied very successfully this concept of liquid structure to interpret: polymerizations of liquids: methyl and higher alkyl methacrylates /64, 66, 72, 73, 87/, propylene /64/, styrene /88/ and styrene modified with nano-silica /89/. Here we shall now deal with the polymerization of methyl methacrylate.

# 7.7. Polymerization of methyl methacrylate

The molecule of methyl methacrylate (MMA), $H_2C=C(CH_3)COOCH_3$ has a double bond (C=C), and can be polymerized by free radical mechanism in same way like ethylene (7-2). Liquid MMA can polymerize at atmospheric pressure and at room temperature up to 90 °C. The polymethylmethacrylate (PMMA) is derived as a material for the production of organic glass (i.e. plexiglass), dental fillings and various other products. It is thought that this simple reaction may be explained by the generally accepted theory of the radical polymerization, which needs to be fast at the beginning of the reaction, and later to slow down as MMA is consumed by the reaction. However, the reaction is not realized just as theory predicts. Initially, the reaction is really fast, and then it slows. But after some time, the reaction begins to accelerate, than it reaches a large value, and then abruptly decelerates and stops completely. This is not in line with theoretical expectations, which also causes some problems in MMA polymerization. Such reaction acceleration is undesirable. It should be prevented. In order to achieve this, we need to know its cause.

Since 1930, a dozen "theories" were proposed attempting to explain the cause of the acceleration /66/. But, there was no success. In the 1960s Kargin and Kabanov /67/ suggested a hypothesis based on the assumption that the liquid MMA, like any other liquid, is partially an ordered system where some molecules are regularly arranged, and some are not. Therefore, they argued that the theory of polymerization of organized monomers systems should be used to interpret the MMA polymerization. However, this hypothesis has never been checked.

Sasuga and Takehisa /69/ have experimentally confirmed that the liquid MMA is actually composed of domains in which the molecules are regularly arranged, and domains of randomly arranged molecules. Korolev and al /70/ have confirmed this for the liquid MMA, and also for 8000 other liquids; the existence of ordered and disordered domains that consists of molecular dimers and linear oligomers.

Based on these and many other studies it can be concluded that the liquid MMA is a partially organized system, but no study has proposed a method to determine the fractions (percentage) of molecules that are in ordered and disordered domains. This can, however, be very easily achieved using Boscovich's concept of liquid structure (Figure 7-7).

There is empirical data for specific volume of the liquid MMA ($V_t$): it is $V_{25°C}=106.7$ cm$^3$/mol at 25 °C and increases up to $V_{75°C}=113.9$ cm$^3$/mol at 75 °C.

Knowing that the critical volume of MMA is $V_c=315$ cm$^3$/mol, covolume b and volume $V_{t,s}$ can easily be calculated by the staircase model in Figure 7-2. Hence, b = 157.5 cm$^3$/mol and $V_{t,s}$ =99.2 cm$^3$/mol.

Specific volume of liquid MMA $V_t$ is less than b, but greater than $V_{t,s}$ i.e. ($b>V_t>V_{t,s}$), (Figure 7-2). In the theory of Savich and Kashanin there is an interval of abrupt change in density (Figure 6-1) and this corresponds to the repulsive arch in Boscovich's curve for fluids (Figure 7-7). Molecules can not exist at that distance; they have to be either on the lower non-cohesion limit or on the higher cohesion limit. Therefore, it is reasonable to assume that the liquid MMA is a mixture of two micro phases: disordered "gas-like" phase consisting of 3D-rotating bimolecules with specific volume that corresponds to covolume b, and ordered "crystal-like" phase consisting of 1D-rotating linear oligomolecules with specific volume that corresponds to $V_{t,s}$. The specific volume $V_t$ of liquid MMA, as a mixture of two micro phases, depends on the fractions (percentages) of these phases according to equations (7-3).

$$V_t = Y_b \cdot b + (100-Y_b) \cdot V_{t,s} \tag{7-3a}$$
$$d_t = 1/V_t = X_b \cdot d_b + (100-X_b) \cdot d_{t,s} \tag{7-3b}$$
$$Y_b + Y_{t,s} = 100 \text{ (volume \%)} \tag{7-3c}$$
$$X_b + X_{t,s} = 100 \text{ (mass \%)} \tag{7-3d}$$
$$d_b = 1/b; \quad d_{t,s} = 1/V_{t,s} \tag{7-3e}$$

Here, $Y_b$ and $Y_{t,s}$ are volume fractions, and $X_b + X_{t,s}$ are mass fractions of bimolecules and oligomolecules, respectively; $d_b$ and $d_{t,s}$ are densities of disordered and ordered phases, respectively. The fractions of disordered and ordered micro-phases are the only unknown terms in equations (7-3), and can be easily calculated.

These fractions depend on temperature. By heating, 1D-rotating oligomolecules disintegrate to 3D-bimolecules. Hence, with the increase of temperature the fraction of bimolecular disordered micro phase increases, and fraction of ordered oligomolecular micro phases decreases. Since $b > V_{t,s}$, specific volume of liquid MMA increases with temperature.

Based on the known values of $V_t$, b and $V_{t,s}$, we calculated the fractions of individual phases in liquid MMA (Figure 7-8).

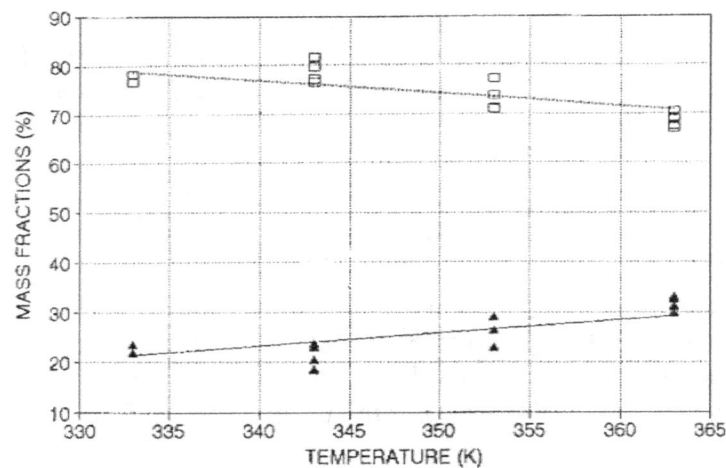

*Figure 7-8. Calculated fractions of ordered (upper line) and disordered phase (lower line) of liquid MMA; experimentally determined fractions of PMMA obtained in ordered (squares) and disordered (triangles) phase /49, 72, 73/*

To test the feasibility of these concepts and the accuracy of the calculations, we polymerized MMA and experimentally determined the fractions of PMMA that have been originated in these phases /49, 72, 73/.

There is a very good agreement between the theoretically calculated and experimentally determined values. Furthermore, many other details concerning mechanism and kinetics of polymerization, as well as the molecular structure of PMMA were interpreted using Boscovich's roadmap.

# 8. PHILOSOPHICAL BASIS OF BOSCOVICH'S COMPREHENSIONS

## 8.1 Introduction

Many authors have discussed Boscovich's philosophical comprehensions /74-77/. In the literature, there are a number of articles about Boscovich's epistemology, philosophy of mathematics, force and matter, space and time. Since there are very thoroughly processed philosophical views on Boscovich, the reader is referred to the literature cited.

Here we want to consider a more detailed look at Boscovich's understanding of forces of attractions and repulsions, which play an essential role in the behaviour of materials. It is not only him, but many of his predecessors, contemporaries and subsequent thinkers have had a similar understanding: Leucippus, Democritus, Heraclites, Aristotle, Empedocles, Toland, Holbach, Newton, Kant, Hegel, Engels... Some of them (Kant, Hegel and Engels) believed that the **attraction and repulsion are the essence of matter**. We emphasis again - **the essence of matter!** Therefore it is astonishing that almost none of our contemporaries, philosophers and naturalists, paid attention to it. Hegel and Engels works, written in 19th century, were widely read in 20th century by philosophers and scientists, which wrote plenty of articles and books about Hegel's and Engels' philosophy. We have read hundreds of these articles and books, but we did not find a single text that even mentioned something about Hegel's and Engels' comprehensions of attraction and repulsion.

Dealing with these aforementioned big thinkers, from the ancient Greeks onwards, about what they thought in terms of attraction and repulsion would be too broad and beyond the scope of this book. Here we shell consider and compare only the concepts of Boscovich, Hegel and Engels in terms of attraction and repulsion.

## 8.2. Attraction and repulsion
### – Comprehensions of Boscovich, Hegel and Engels /78/

> *Hegel was quite right in saying that the essence of matter is attraction and repulsion.*
> *(Engels, "Dialectics of Nature")*

The very fact that the great thinkers Boscovich, Hegel and Engels devoted so much attention to attraction and repulsion suggests that the issue is important. Therefore, our goal is to underline their basic understandings about the meaning of attraction and repulsion. We will show the importance of the analysis of attraction and repulsion for the interpretation of certain phenomena in nature.

## 8.2.1. Boscovich's conception of the attraction and repulsion

Boscovich's Theory of natural philosophy /8/ was based on attraction and repulsion. He believed that the basic elements of matter are non-extended and indivisible points, scattered in the infinite vacuum. The distance between points of matter can be infinitely increased or decreased, but it can not completely disappear.

If two points approach each other, then, there is a cause that leads to a deceleration or acceleration of their movements. At certain distances two points are determined to be approaching, and at other distances they move away. That cause, which changes the state of the body concerning its motion and state of rest /8/, i.e. this determination to move away or towards is named by Boscovich as repulsive or attractive forces. The law of the force is such that repulsion and attraction alternate as the points approach each other (known as Boscovich's curve, Figure 3-1).

When points come to a negligible distance, repulsive force is infinitely large and can destroy any chance of them meeting, no matter how large the speed at which a point comes closer to the other point; making it impossible for the distance between the points to completely disappear.

And as the points of matter move away from each other, the repulsive force reduces, until at greater distance it becomes attractive, and then farther out the force again becomes repulsive and so on alternatively, until it becomes permanently attractive at a large mutual distance between the points.

According to Boscovich, attraction and repulsion are forces. However, it is wrong to understand the attraction and repulsion as two types of forces. "Both kinds of force belong to the same species; for one is negative with regards to the other, and a negative does not differ in species from positives. That the one is negative with regard to the other is evident from the fact that they only differ in direction, the direction of one being exactly opposite of the direction of the other; for in the one there is a propensity to approach, in the other a propensity to recede... /8, Section 108/."

The result of the action of these forces is a movement comprised of approaches inward or recessions outward. Since the quantity of motion in the Universe is maintained as always the same, Boscovich indicates the sum of all the attractive movements is equal to the sum of all recessive movements in each moment /8, Sections 261, 264/.

By Boscovich, transformation of attraction to repulsion and vice versa is possible, and fulfilled at so-called limits of cohesion and non-cohesion (Chapter 3.1.).

## 8.2.2. Hegel's conception of the attraction and repulsion

For Hegel, **rrepulsion is the fragmentation of one into the many ones /79, 80/. This is a negative reference of the one to itself**. This repulsion generates many ones and it enables the existence of ones as the one.

For old Greek atomists, the matter consists of atoms and void. Hegel concludes that the void, which is assumed as the complementary principle to the atoms, is repulsion and nothing else, presented under the image of the nothing existing between the atoms. (See comment at the end of chapter 8.2.4)

The ideality will, however, be realized in attraction. **This self-positing-in-a-one of the many ones is attraction**. Repulsion passes over into attraction, the many ones into one. Both, repulsion and attraction are at first distinguished from each other, repulsion as the reality of the ones, attraction as their posited ideality.

Although negative, repulsion is nonetheless essentially connection. Attraction refers to repulsion by having it for a presupposition. Repulsion delivers the material for attraction. If there were no repelled ones, there would be nothing to attract.

The one is, however, ideality that has been realized, posited in the one; it attracts through the mediation of repulsion; it contains in itself this mediation as its determination. It thus does not swallow the attracted ones within it as into one point. Since it contains repulsion in its determination, the latter equally preserves the ones as many within it.

As thus determined, they (i.e. attraction and repulsion) are inseparable... Thus, repulsion is the positing of the many; attraction the positing of the one.

Hegel noticed that attraction and repulsion, as is well known, are usually regarded as forces. He does not agree with the practice in natural science to explain the phenomenon with forces. The nature of force itself is unknown and only its manifestation apprehended. Hence, the explanation of a phenomenon by a force is a mere tautology.

He noticed that Kant famously constructed matter from the forces of repulsion and attraction. "Now even if such a so-called construction of matter had at most analytical merit, however diminished because of a flawed exposition, the thought on which it is based, namely that matter must be made out to be from these two opposing determinations as its fundamental forces, must always be highly esteemed."

## 8.2.3. Engels' comprehension of attraction and repulsion

Engels' comprehension of attraction and repulsion is outlined in his unfinished work "The Dialectics of Nature" /81/, which is made up of articles, scraps and fragments written in the period from the 1873 up to 1886 (Table 8-1). Attitudes about the attraction and repulsion are woven into the whole work and represent the basis for Engels' dialectics of nature.

Engels started from an attitude that motion, in the most general sense, is conceived as the mode of existence of matter, comprehending all changes and processes in the universe, from mere change of place right up to the process of thinking /81, p. 74/. There is no matter without motion, and there is no motion without matter. The matter as well as the motion can not be created nor destroyed.

Table 8-1. Attraction and repulsion in Engels' "Dialectics of Nature" /81/

|  | Chapters and issues interpreted by attraction and repulsion | Page | Year of writing |
|---|---|---|---|
| 1. | **Outline of the part plan**<br>- Transfer of motion, the conservation of energy law | 17 | 1880 |
| 2. | **Basic forms of motion**<br>- The interaction of two bodies<br>- The planet's rotation<br>- Earth mechanics<br>- Heat<br>- Electricity and magnetism<br>- Chemical processes<br>- Changing one form of motion to another<br>- The importance of Solar energy for processes on Earth<br>- The concept of force<br>- Origin of the Solar system | 74-93<br>76-78<br>78<br>80<br>82<br>83<br>83<br>84<br>86<br>87-93<br>90-93<br>90-93 | 1880-81 |
| 3. | **Dialectics. General questions of dialectics. The fundamental laws of dialectics.**<br>- Magnetism, electricity, chemical processes | 239 | 1875 |
| 4. | **Forms of motion of matter. Classification of the sciences.**<br>- General concepts of attraction and repulsion<br>- Repulsion in the tails of comets and gas<br>- Attraction and gravitation<br>- Dissipation and condensation of matter<br>- Transformation of attraction into repulsion and vice versa<br>- Origin of the Solar system<br>- Thermal expansion and repulsion<br>- Motion and equilibrium<br>- Differentiation of matter<br>- Motion of the heavenly bodies<br>- Motion on one heavenly body<br>- Conversion of one form of motion into another | 275<br>275<br>276<br>276<br>276<br>277<br>277<br>278-279<br>278-279<br>179-280<br>279-280<br>280<br>279-280<br>280 | <br><br><br><br>1874<br>1874<br>1874<br>1874<br>1874<br>1880<br>1880<br>1880<br>1880<br>1880 |
| 5. | **Physics**<br>- The concept of force<br>- Repulsion is active, attraction is passive<br>- Transformation of attraction to repulsion in gases | 320<br>324<br>325 | 1880<br>1880<br>1873 |

The movement of each material carrier (particle or body) is bound up with some change of place. This change of place can consist only in coming together or separation. "Hence the basic form of all motion is approximation [*see note*] and separation, contraction and expansion - in short, the old polar opposites of *attraction* and *repulsion*". According to Engels, **it is expressly to be noted that attraction and repulsion are not regarded here as so-called "forces" but as *simple forms of motion* – approximation [see note] and separation.** (And his comprehension of the forces Engels outlined later in the "Dialectic of nature" and we will look at it.) [*Note of translator: It is more appropriate to say "getting closer together" than "approximation."*]

"All motion consists in the interplay of attraction and repulsion. Motion, however, is only possible when each individual attraction is compensated by a corresponding repulsion somewhere else. Otherwise in time one side would get the preponderance over the other and the motion would finally cease. Hence all attractions and all repulsions in the universe must mutually balance one another. Thus the law of indestructibility and uncreatability of motion is expressed in the form that each movement of attraction in the universe must have as its complement an equivalent movement of repulsion and vice versa; or, as earlier philosophy – long before the natural-scientific formulation of the law of conservation of force or energy – expressed it: the sum of all attractions in the universe is equal to the sum of all repulsions" /81, p. 77/.

Furthermore, Engels excluded the possibility that these two opposites mutually cancel or separate. He points out that for dialectical comprehension based on natural sciences, it is impossible that any two opposites completely separate or annul each other.

On the examples of the rotation of the planets around the sun, lifting and falling bodies, movement of molecules, magnetism and electricity, Engels concludes that the form of motion here conceived as **repulsion** is the same as that which modern physics terms **energy**, and an expression of **force** is what in physics mistakenly used instead of using the term **attraction**. By the example of the chemical reaction of oxygen and hydrogen, Engels shows the importance of the liberation of heat and concludes: "In the overwhelming majority of cases, motion is given off on combination and must be supplied on decomposition. Here, too, as a rule, repulsion is the active side of the process more endowed with motion or requiring the addition of motion, while attraction is the passive side producing a surplus of motion and giving off motion" /81, p. 83/.

Engels paid a lot of attention to the transformation of one form of motion to another. He concluded that it is essentially a transformation of one form of repulsion to another form. Thus, when a falling body hits the ground, the movement of body has been converted to a lesser extent into the air vibrations of sound waves, and to much greater extent into heat. The repulsion of masses (body – Earth) is transformed into molecular repulsion.

As stated above, it can be seen that Engels conceives repulsion as motion which comprises stretching, separation, alienation. In order to achieve this kind of motion it is necessary to bring energy. The attraction is also a form of motion, which is the opposite expression to repulsion, i.e. approaching, attracting, coupling, compression. In order to achieve this kind of movement it is necessary take away the energy.

Engels opposed the opinion that gravity is the most general determination of materiality, because this would then mean that attraction is a necessary property of matter, but not repulsion (as a necessary property of matter), since gravity being considered only attractive.

"But the attraction and repulsion are as inseparable as positive and negative, and hence from dialectics itself it can already be predicted that the true theory of matter must assign as important a place to repulsion as to attraction and that a theory of matter based on mere attraction is false, inadequate, and one-sided... The whole theory of gravitation rests on saying that attraction is the essence of matter. It is necessarily false. Where there is attraction, it must be complemented by repulsion. Hence, already Hegel was quite right in saying that the essence of matter is attraction *and* repulsion. And, in fact, we are more and more becoming forced to recognize that dissipation of matter has a limit where attraction is transformed into repulsion, and conversely the condensation of the repelled matter has a limit where repulsion becomes attraction" /81, p. 276/.

Transformation of attraction into repulsion and vice versa is possible, says Engels, but Hegel explains it in a mystical way, but also in fact in the correct way. Engels emphasises that "Hegel shows his genius even in the fact that derives attraction as something secondary from repulsion as something preceding it." To illustrate, Engels points out that "a Solar system is only formed by the gradual preponderance of attraction over the originally prevailing repulsion."

Considering the attraction only as a form of (approaching) motion Engels points out that the attraction should not be conceived as a force. Namely, Engels believes that the notion of force is always used for the interpretation of phenomena whose causes we have not been able to find out. Thus, it was spoken and is still spoken of things like magnetic force, electric force, capillary forces, chemical forces, vital force, gravitational force, electric contact force on metals etc. "So, **force** here reaches its limit... Here it becomes a phrase, as everywhere where, instead of investigating the uninvestigated forms of motion, one *invents* a so-called **force** for their explanation..., in which case as many forces are obtained as there are unexplained phenomena, the external phenomenon being indeed merely translated into an internal phrase" /81, p. 320/. But, as science has progressed and one by one of these occurrences was explained, one by one these forces has disappeared. It was always found that there was not a force but a transformation of one form of motion to another.

"In order to understand the separate phenomena we have to tear them out of the general inter-connection, and consider them in isolation. and *then* the changing motions appear, one as cause and the other as effect... We cannot go back further than to knowledge of this reciprocal action, for the very reason that there is nothing behind to know. If we know the forms of motion of matter (for which it is true there is still very much lacking, in view of the short time that natural science has existed), then we know matter itself, an therewith our knowledge is complete" /81, p. 262 /.

According to Engels, the notion that the force is the cause of motion is transferred from mechanics. "In mechanics the causes of motion are taken as given and their origin is disregarded, only their effects being taken into account. Hence, if a cause of motion is termed a force, this does no damage to mechanics as such; but it becomes the custom to transfer this term also to physics, chemistry, and biology, and then confusion is inevitable" /81, p. 93 /. "And finally, in every natural science, even in mechanics, it is always an advance if the word *force* can somewhere be got rid of" /81, p. 177/.

## 8.2.4. Distinctions and similarities in conceptions of Boscovich, Hegel and Engels

Let's look at what are the differences in the conceptions of attraction and repulsion by Boscovich, Hegel and Engels.

By Boscovich, attraction and repulsion are forces and as such, they are determination or propensity to attract and repel.

For Hegel, the attraction is the setting of many ones into the unique one, that is, the creation of a complex particle by assembling and connecting the simpler ones. Repulsion is the fragmentation of the one into the many ones, the denial of their mutual relationships, therefore, degradation of complex particles into simple ones.

The motion, which is essential for achieving the attractions and repulsions, Hegel called "attracting" and "repelling" and therefore distinguished that motion from the attraction and repulsion itself. For Engels, however, it is just these motions, the move closer together or removal (move further apart), denoted as attraction or repulsion.

However, despite these differences, there are many similarities in their attitudes. First of all, they have the common approach to interpretation of the essence of matter through the dialectical unity and struggle of two opposites – attraction and repulsion. They agree that the attraction is inseparable from the repulsion and that each attraction is compensated by an equally valuable repulsion. They agree that the attraction can transform into repulsion, and opposite, although they do not explain in more detail the way in which this transformation is realized. According to Engels it is possible to convert repulsion of any quality to repulsion of another quality. This Engels interprets as converting one form of energy into another.

All three consider the issue of limits of attractions and repulsions. If the attraction implies the coming together of material particles (Engels' understanding of the concept of attraction), then Boscovich, Hegel and Engels agree that this attraction is limited.

Separation of material particles (Engels understanding of the concept of repulsions) is also limited by Engels comprehensions, not limited by Boscovich, while Hegel does not clarify the issue. If, however, the attraction involves **force** (Boscovich), then at infinitely short distances, this force vanishes, and at very great distances it exists, but has a small value (gravity). At any other distance, the attraction force can be of arbitrary value (from zero to infinity). Regarding repulsion, according to Boscovich, it is infinitely large at the infinitely small distances, there is none at infinitely large distances, and this interval may have any arbitrary value.

Hegel concludes that the void, which is assumed by old Greek atomists as the complementary principle to the atoms, is repulsion. Hence,

|          |        |            |
|----------|--------|------------|
| By Hegel: | Void | = Repulsion |
| By Engels: | Energy | = Repulsion. |
| Consequently: | Void | = Energy |

It means that equal void in matter corresponds to its equal energy! Is it true? Let's look some examples:

(1) **Potential mechanical energy ($E_p$) of a body** is $E_p$=mgh, or calculated per unit of mass (m), $E_p/m$=gh. It means, equal distance (h) from the ground, i.e. void body to Earth – equal energy per unit of mass body, regardless the chemical nature of matter.

(2) **Energy of ideal gas** per one mole is E=PV=RT. (P is pressure, V is molar volume, T is absolute temperature, and R is universal gas constant.) By Avogadro there is equal number of molecules in equal volumes of gases, i.e. by Ampere the average distance between molecules in gases is equal, regardless the chemical nature of gases. It means equal void between molecules in gases – equal energy of gas.

(3) **Nuclear energy** calculated per one nucleon approximately 8 MeV and is equal for all heavy nuclei having atomic mass above 16. These nuclei have equal density, i.e. the void between nucleons is equal. It means equal void between nucleons – equal energy.

On three levels of hierarchy of matter, i.e. mechanical, molecular and nuclear, it is evident that comprehensions of Hegel and Engels are correct.

## 8.2.5. Analysis of Boscovich's comprehension of attractive and repulsive forces

By the comprehension that attraction and repulsion are only forms of motion and by opposition to the interpretation of phenomena in the natural sciences that views this as some force, Engels, and Hegel oppose Boscovich, because his entire theory of matter is devised on attractive and repulsive forces. Who is right?

We already wrote about the examples that confirm the correctness of Engels' conception of attraction and repulsion /45, 55/. Also we have provided examples that demonstrate the validity of Boscovich's comprehensions (Chapter 5). From this it follows that Hegel and Engels are right to reject the use of the concept of **forces**, and Boscovich just uses that concept. Before we explain this contradiction, let's look at some problems arising from the use of force by Boscovich.

Attraction and repulsion are understood by Boscovich as **forces** and represent determining whether particles could approach or recede. By Boscovich the exertion of these forces consists also in the acceleration (or deceleration) of mutual movement of particles. By Boscovich, the first part of his curve is repulsive force which is able to destroy any speed, no matter how great it is.

So we conclude that from the actual motion there remains nothing but some determination of the motion, as a possibility. This determination to approach or to recede, in certain conditions is realized in actual motion, namely as the acceleration (or deceleration) of existing motion. Thus, therefore, force at least seemingly, creates and destroys motion.

However, Boscovich, Hegel and Engels accepted Descartes' opinion that the quantity of motion in the universe remains constant, i.e. Descartes' principle of conservation of motion. Consequently, Engels concluded that the total motion can not be created or destroyed.

Evidently, there are some contradictions between the principle of conservation of motion and ability of force to create and destroy motion.

Hence, we conclude that force is basically the realization of the transformation of one form of motion into another form of motion. We conclude that the action of force only seems as a creation or destruction of movement, but is rather some qualitative and quantitative change of the form of motion. (This notion of force and its manifestation is emphasized even by Engels, as we have just written.)

Natural science teaches us what are the forms of motion that are complementary and mutually interchanged and, therefore, converted into one another. It is known, for example, when two molecules in a gas fly towards each other, their approach and collision (attraction by Engels) is accompanied by repulsion within themselves – initially, the electrons move away from the atomic nuclei (i.e. electronic polarization of molecules), and then the electrons again move towards the nuclei and in each molecules simultaneously the atoms move away from one another (atomic polarization) /82/, and the decomposition of molecules is possible (i.e. chemical reaction).

These movements and changes inside the molecules affect the slowing down or speeding up the mutual movement of molecules as the whole. Therefore, in modern science the repulsive and attractive arches, which are in Boscovich's curve, are interpreted as disturbance of balance in the particles as they come closer.

**While at one level of the structure of matter there predominates attraction** (the two molecules coming closer), **on another (lower) level predominates repulsion**, i.e. electrons inside the molecules move away from the nuclei. And when the molecule begins to decompose into atoms (i.e. atomic repulsion), simultaneously this new type of repulsion is compensated by attraction of electrons toward individual nuclei. So, attractive and repulsive forces acting between the molecules are the transformation of one form of motion to another, inter-molecular to intra-molecular and vice versa. Here, the force arises as the relation of the external and internal motions, and "by the manifestation of force the inward is put into existence" (Hegel). Accordingly, the forces of attraction and repulsion, as well as the arches on the Boscovich's curve, are the reflection and measure of the particles distortion (fighting repulsion and attraction among themselves), which occurs as a result of particles approaching each other.

Boscovich (and Hegel and Engels) talks about transformation of attraction into repulsion and vice versa. We believe that such a conversion is not possible. Actually, we wonder ourselves: what happens to the force of attraction when two points come to a distance at which they are determined to be repelled? Does that force of attraction disappear, and be destroyed? Is not then, at least for a moment, in no matter how small a space, the world poorer by one force of attraction? Is it not in that moment violated the principle, which all three philosophers emphasize, that at all times the sum of all attractions equal the sum of all repulsions?

We believe that the transformation of attraction to repulsion (and vice versa) is only apparent: in fact the attraction of one form is converted into attraction of another form. Repulsion is also converted into another form of repulsion. It confirms the previous example of the approach and collision between two molecules. This attraction occurs at the expense of reduced attraction inside the molecules. The original repulsion between the two molecules is decreased, as it is simultaneously converted into repulsion of their constituent parts (i.e. electrons and atoms). Thus, through the unity and struggle of two opposites the original state is negated and a new state is created in which both opposites are present but at qualitatively other levels.

### 8.2.6. Differentiation of matter

Although the terms of attraction and repulsion have different meanings in the comprehensions of Boscovich, Hegel and Engels, all three meanings are related to the different consecutive phases (steps) of the same process – **differentiation of matter**. Differentiation of matter means a creation of a difference in the homogeneity, i.e. breakdown of a whole that consists of the same type of particles into parts of different types of particles.

For example, in the cloud of water vapour the molecules are separated from one another and individually move in different directions. They are the same type. By condensation, drops of water are formed. The creation of the first drop is the emergence of new particles and new type of movement, i.e. the difference is created from the same type of molecules. This is differentiation of matter – drop of water is differentiated, created, and springs into being. **The gathering of many individual molecules in one drop of water** – that is the attraction by Hegel. ("The self-positing-in-a-one of the many ones is attraction.") And to be together, these molecules need to sacrifice the freedom of independent random movement and to come closer to each other until a drop of water is formed. This coming together is attraction, according to Engels. Simultaneously with the approach of molecules towards each other, the atoms in them are going away from each other and hydrogen bonds are created among molecules in a drop. This distortion of the molecules, the vanished external and into internal transformed motion, the hydrogen bonds and molecular **forces** creation is nothing else, but attraction spoken of by Boscovich.

By formation of water drops, molecules come into equilibrium in which seemingly pervades attraction over repulsions. "The possibility of particles being at relative rest, the possibility of temporary states of equilibrium, is the essential condition for the differentiation of matter" (Engels /81, p. 278/). But there is no absolute equilibrium, "all equilibrium is only relative and temporary". The equilibrium can not be separated from movement; the individual motion strives toward equilibrium, "the motion as a whole once more destroys the individual equilibrium". In fact, the movement of a drop of water is a new type of movement, which had not existed before the water drop was formed. This movement of drops of water (to one side), and the movement inside the molecules (to the other side) are the result of the vanished movement of the individual molecules of water.

We see that the **differentiation of matter is simultaneously the differentiation of movement** – it emerges the new and arouses the old forms of movement that have been surpassed in the previous stage of differentiation. Thanks to this, the opposition of attraction and repulsion (in both Hegel and Engels and Boscovich meanings) is not annihilated, but only altered into other forms of the two opposites. The new one is created (drop of water), that exists due to repulsions from other same things (other drops) (Hegel's sense repulsions) and now we can re-watch the game of attraction and repulsion (in all three senses), but this time we no more look at the molecules, but look at the drops of water. If there was no differentiation of movement on the higher and lower levels, attraction and repulsions would be simple, mechanical assembly and disassembly, gathering and pulverization, without change the quality of material. Since the differentiation of matter necessarily involves the differentiation of movement, the movement is not a mere change of place as it is in area of mechanics, but out of it – is change of quality.

Common to all three thinkers is the statement that total sum of all attractions and sum of all repulsions in the universe is always constant (**the law of conservation of attraction and repulsion**), and it is attained since each individual attraction has to be compensated by some repulsion of same value. It is possible that attraction increases and repulsion decreases at some level (n) in the structure of matter. It could seem that above mentioned law is violated.

But, it is not violated. By examples taken from modern science we have shown that the opposite processes occur simultaneously both on the higher (n+1) and lower (n-1) levels, i.e. an decrease of attraction and an increase of repulsion /45, 53, 78/. And contrary – each increase of repulsion at some level of structure is compensated by an adequate increase of attractions at the higher and at the lower levels (Table 8-2). Empedocles (5th century BC) was pointed to this rule by the statement that love combines and strife separates: when the whole is separated into parts, their elements in are combined. Contrary, when two bodies are combined, their elements are separated.

Table 8-2. An increase (↑) of some opposite at level (n) is compensated
by its decrease (↓) at higher (n+1) and lower (n-1) levels

| Level | Attraction and Repulsion | Order and Disorder | Homogeneity and Inhomogeneity | Necessity and Chance | Continuity and Discontinuity |
|---|---|---|---|---|---|
| n+1 | ↑R, A↓ | ↑D, O↓ | ↑I, H↓ | ↑C, N↓ | ↑D, C↓ |
| n | ↑A, R↓ | ↑O, D↓ | ↑H, I↓ | ↑N, C↓ | ↑C, D↓ |
| n-1 | ↑R, A↓ | ↑D, O↓ | ↑I, H↓ | ↑C, N↓ | ↑D, C↓ |
| References | 45, 53, 78 | 45, 53, | 53 | 53, 64 | 25, 53, 62 |

By analyzing the achievements of modern physics and chemistry, we have shown /25, 45, 53, 62, 78/ that some other opposites (e. g. order-disorder, homogeneity-inhomoheneity, necessity-chance, and continuity-discontinuity) were altered and mutually compensated on the same manner as the consequences of that Empedocle's rule. For example, an increase of order of some level is compensated by an increase of disorder both on the higher and lower levels. This extension and deepening of philosophical comprehensions of Boscovich, Hegel and Engels are of a great importance for understanding of modern scientific theories, and for their application, too.

# 9. ROGER BOSCOVICH – THE FOUNDER OF MODERN SCIENCE

## 9.1. Influence of Boscovich's Theory on the contemporaries and followers

Boscovich's philosophical and natural-scientific concepts were known to many of his contemporaries and subsequent scientists who declared themselves as supporters and followers of at least some of his comprehension. In fact, in 18th and 19th century his Theory and other works were in the curricula in many universities and educational institutions in cities that are now within Austria, Italy, France, Germany, England, Poland, Hungary, Croatia... Also, his Theory was in contents of many books and encyclopaedias in the 18th and 19th century /2/. For example, White /6a, p. 104/ highlights that famous "Encyclopaedia Britannica" from 1801 contained an article of fourteen pages about Boscovich and his Theory. Therefore, there should have been a lot of people at that time familiar with Boscovich's Theory from this very popular encyclopaedia.

Many scientists, whose names are familiar to us, were inspired by his Theory: Ampere, Cauchy, Fehner, Priestley, Gay Lisak, Faraday, J. J. Thomson, W. Thomson (Lord Kelvin), Mendeleyev, Helmholtz, Hertz, Maxwell, Lorentz, Davy, Bohr, Heisenberg and many others, which is detailed in the literature /1-7/. Here we mention just some opinions of prominent scientists about the role and importance of Boscovich's Theory in development and emergence of modern science.

In the opinion of Zh. Dadich /2, p. 98/ the best value Boscovich's Theory is as Zh. Markovich appreciated /1/: "What is Boscovich's essential contribution in his Theory of natural philosophy? It is the building of a qualitative scheme for representing mechanical as well as other properties of matter leading to a completely new view on the composition of matter. The emphasis is not a more or less good quantitative representation of observed phenomena but on the immense possibilities of variation in the application of the scheme."

German physicist Werner Heisenberg, now known for the principle of uncertainty, in 1958 wrote: "He (namely Boscovich) regarded matter as a space filled with a field of force in which elementary particles represent so to speak only single points in the field. The concept of the field of force, which played such a decisive role in the development of physics as of the 19th century is found already by Boscovich, later leading to fundamental developments, e.g. Faraday's work" /2, p. 116/.

Lord Kelvin very critically evaluated Boscovich's Theory, sometimes accepting and sometimes rejecting it, and finally in 1907 was quite sure that Boscovich's Theory could be applied to explain phenomena in the interior of the atom, and declared: "My present assumption is Boscovichianism pure and simple" /2, p. 122/.

In 1958 Niels Bohr pointed out that Boscovich had a large role in the creation of modern science. He points out that Roger Boscovich, whose life's work is gaining increasing attention in the scientific world today, was one of the most prominent figures in 18th century among philosophers of nature who enthusiastically developed core of Newtonian mechanics ideas. "Indeed, he did not only make important contributions to mathematics and astronomy, but strove with remarkable imagination and logical power to develop a systematic account of the properties of matter on the basis of interactions of mass points through central forces. In this respect **Boscovich's ideas exerted a deep influence on the work of the next generation of physicists**, resulting in the general mechanistic views which inspired Laplace and, perhaps less directly, even Faraday and Maxwell" /5, p. 8 and 184/.

The great Russian chemist D. Mendeleyev (1834-1907) in his classic book "Fundamentals of Chemistry" presents Boscovich's Theory, and says that Boscovich was the founder of modern atomism and marks him with Copernicus as the pride of the Western Slavs.

Interestingly, a similar thought is given by philosopher F. Nietzsche who said that Boscovich with Copernicus was so far "the largest and the most victorious opponent to illusion" and that his work was the biggest triumph over the senses that has so far been made on Earth /1, pp. 469 /.

## 9.2. Resurrection of Boscovich's Theory

We've travelled over a long tour in the hierarchy of matter – from elementary particles, nucleons, atoms, molecules, supra-molecular particles, macromolecules, nano-particles, colloids, and up to the planets in the Solar system. On this tour, we've always encountered Boscovich's traces, marks and confirmations of his comprehensions. We have showed that many stages of scientific knowledge are strewed by Boscovich's traces.

The examples presented here, as well as many others that were not mentioned, testify to the correctness of Boscovich's comprehension of nature. But, we do not want to leave the reader believing that Roger Boscovich, as resident of 18th century, was only a visionary, a forerunner or just a prophet of future scientific discoveries. If Boscovich's work is understood that way, then an injustice would be made. He's not a prophet nor visionary, but the founder of many future discoveries. Namely, as it was said, his Theory was taught in many universities and educational institutions in cities across Europe. Boscovich's Theory had been in the content of many books and encyclopaedias during the 18th and 19 century. His Theory has inspired many famous scientists, who built the magnificent edifice of modern science - built on **the foundations of the Theory of Boscovich**.

Hence, it is no wonder that we are again faced with the concepts that are similar to Boscovich's: his law of force was an **initial assumption** (hypothesis) for the interpretation of the structure of matter, and when the structure of matter was perceived (as **a consequence of** the initial assumptions) and when the law of force acting between the particles was theoretically and **experimentally** determined, it is quite understandable that it **confirmed the initial assumption**, i.e. Boscovich's law of force. This is a logical way to do modern research: it is started by some hypothesis-assumption that has to be proved, and it is proved by performing of some logical consequences of this assumption, than that consequences are experimentally verified and then it is concluded whether the experimental results are consistent with the initial assumption. If so, the initial hypothesis was proven. Today, science is so developed that that path (initial assumption → logical consequences → experimental confirmation → proven theory) takes only several years.

However, the path from Boscovich's Theory to the construction of modern science had not been completed at once, overnight. Founded by Boscovich in 18th century, it was constructed and discarded, many times, like in the poem "The Building of Scadar on Bojana", continued in the 19th and largely completed in 20 century; many generations of scientists participated in that building, and it is simply forgotten what were the initial foundations and who started them. Therefore, through mediation of all these scientists, Boscovich's comprehensions reach to us as an echo of the past, often leading us to think by Boscovich's minds, but we are not aware about that.

For these reasons, one should respect the opinion of the famous W. Heisenberg who pronounced in 1958: "His main work, 'Theoria Philosophiae Naturalis', contains numerous ideas which have reached full expression only in modern physics of the past fifty years, and which show how correct were the philosophical views which guided Boscovich in his studies in the natural sciences" /2, p. 126/.

Interestingly is the opinion of Lederman that Boscovich was "ahead of his time": "His Theory was incomplete and limited, but the idea of particles having a radius equal to zero and looking like a point, and yet produces around itself a 'force field' **is the key to the whole of modern physics**" /83, p.108 and 345/. [*Translator's note: Citation is based on translation from Serbian edition.*]

"Boscovich's masterpiece has gained more and more attention in the scientific world nowadays, and Frenchman Herrismann even claims that Boscovich's whole philosophy of nature will become the philosophy of the next (21st) century" (V. Filipovich in Epilogue of /8, Liber edition/).

Roger Boscovich was a great thinker and scholar of his time. The founder of modern scientific laws of the micro and macro world, Boscovich's bright mind embodied in the Theory, can give to the modern scholars incentive to recognize and perceive the world around them, to recognize the meaning of the modern science achievements... and to continue to work and build a new world based on natural laws and universal values.

Our message in the end is: Introduce yourself with Boscovich's Theory in order to recognize it in the achievements of modern science and – **know how you could apply it**!

# REFERENCES

/1/ Marković Ž., "Ruđe Bošković", Jugoslavenska akademija znanosti i umjetnosti, Zagreb, 1968. (part one) i 1969. (part two).

/2/ Dadić Ž., "Ruđer Bošković", Školska knjiga, Zagreb, (bilingual: English and Croatian), 1987. (first edition), 1998. (third edition)

/3/ Stipanić E., "Ruđer Bošković", Dečje novine, Gornji Milanovac, Prosvetni pregled, Beograd, 1984.

/4/ Dimitrić R., "Ruđer Bošković", Helios, Pitsburg-Beograd, 2006.

/5/ Supek I., "Ruđer Bošković - Vizionar u prijelomima filozofije, znanosti i društva", Školska knjiga, Zagreb, 2008.

/6/ (a) White L. L., editor, "Roger Joseph Boscovich - Study of his life and work on the 250th anniversary of his birth", George Allen and Unwin, London, 1961.; (b) White L. L., "Boscovich and particle theory", Nature, **179**, 284 (1957)

/7/ Gill H. V., "Roger Boscovich, S. J. - Forerunner of modern physical theories", M. H. Gill and Son, Ltd., Dublin, 1941.

/8/ Bošković R., "Philosophiae naturalis theoria redacta ad unicam legem virium in natura existentium", Beč, 1758. (first edition), Venice, 1763 (second edition); "A Theory of natural philosophy", The Massachusetts institute of technology, M.I.T. Press, Cambridge, 1922. i 1966.; "Teorija prirodne filozofije svedena na jedan jedini zakon sila koje postoje u prirodi", (bilingual: Latinski i Croatian), Liber, Zagreb, 1974.

/9/ Stoiljković D., (a) "Ruđer Bošković - astronom, fizičar i matematičar", Zbornik radova konferencije "Razvoj astronomije kod Srba IV", Beograd, 22.-26. april 2006., str. 217-225., (b) Stoiljković D., editor, "Ruđer Bošković", Gradac, **180-181**, 1-260 (2011)

/10/ Bošković R., "O zakonu kontinuiteta i njegovim posledicama u odnosu na osnovne elemente materije i njihove sile", Matematički institut, Beograd, 1975.

/11/ Bošković R., "O zakonu neprekinutosti", Školska knjiga, Zagreb, 1996.

/12/ Bošković R., "Pomračenja Sunca i Meseca", Astronomsko društvo "Ruđer Bošković", Beograd, 1995.

/13/ Bošković R. J., "Dnevnik sa puta iz Carigrada u Poljsku (1762. godine)", izdanje "Rajković", Beograd, 1937.

/14/ Stoiljković D., "Teorija Ruđera Boškovića kao putokaz ka kvantnoj mehanici", Arhe, **2**, 181 (2005)

/15/ Stoiljković D., "250 godina kvantne teorije Ruđera Boškovića", Kroz prostor i vreme, br. 3, 41-43 (2009)

/16/ Tomić A., "Lex unica virium in natura - Ruđera Boškovića" u "Epistemološki problemi u nauci", Grujić P. i Ivanović M., editors, Institut za kriminološka i sociološka istraživanja, Beograd, 2004., str. 286-300.

/17/ Stoiljković D., "Putokazi Ruđera Boškovića", Planeta, br. 25, 22-23 (2007)

/18/ Stoiljković D., " Ruđer Bošković - preteča savremenog shvatanja strukture atoma", Hemijski pregled, **49** (3), 54-57 (2008)
/19/ Longair M., "Theoretical concepts in physics (An alternative view of theoretical reasoning in physics)", Cambridge university press, Cambridge, 2nd edition, 2003.
/20/ "Misterija nauke - od čega je sve ovo", Astronomija, br. 23, 34 (2007)
/21/ Diels H., "Predsokratovci - fragmenti", II svezak, Naprijed, Zagreb, 1983.
/22/ Stoiljković D., "Prazni atomi i popunjena praznina", Astronomija, br. 25, 25 (2007)
/23/ Stoiljković D., "Od elementarnih čestica do makromolekula – tragovima Ruđera Boškovića u povodu 225. godišnjice izdanja Boškovićeve Teorije", Polimeri, **4,** 289-291 (1984)
/24/ Stoiljković D., "Makromolekulska hipoteza Ruđera Boškovića", Svet polimera. **9** (5), 225-230 (2006)
/25/ Stoiljković D., "Importance of Boscovich's theory of natural philosophy for polymer science", Polimery, **52**, 804-810 (2007)
/26/ Rinard P. M., "Quarks and Boscovich", Am. J. Phys., **44,** 704 (1976)
/27/ Silbar M. L., "Gluons and Glueballs", Analog, **102,** 52 (1982)
/28/ Benson S. W., "The foundations of chemical kinetics", McGraw-Hill, New York, 1960, str. 213.
/29/ Croxton A. C., "Liquid state physics - A statistical mechanical introduction", Cambridge university press, Cambridge, 1974.
/30/ Portnoy K. I., Bogdanov V. I., Fuks D. L., "Raschet vzaimodeystvii i stabilnosti faz", "Metallurgiya", Moskva, 1981.
/31/ Kaplan I. G., "Vvedenie v teoriyu mezhmolekulyarnykh vzaimodeistvii", Nauka, Moskva, 1982., str. 175.
/32/ Gaylord N. G., Mark H. F., "Linear and stereoregular addition polymers", Interscience, New York, 1959.
/33/ Feng J., Ruckenstein E., "Monte Carlo simulation of interactions between nano-particles", Colloids and Surfaces A: Physicochem. Eng. Aspects, **281**, 254 (2006)
/34/ de Gennes P. G., "Scaling concepts in polymer physics" Cornell university press, Itacha, 1979.
/35/ Đaković Lj. , "Koloidna hemija", Tehnološki fakultet, Novi Sad, 1985.
/36/ Bennett R. H., Hulbert M. H., "Clay microstructure", International human resources development corporation, Boston/Houston/London, 1986., p. 34.
/37/ Keller C., "Radiochemie", Diesterweg Salle, Frankfurt, 1975., Fig. 2.18 (Russian edition: Keller K., "Radiokhimiya", Atomizdat, Moskva, 1978.).
/38/ Migdal A. B., "Fazovie prevrashcheniya yadernogo veshchestva" u E. B. Etingof, "Nauka i chelovechestvo", Znanie, Moskva, 1978., str. 138-147.
/39/ Grigor'ev V., Myakishev G., "Sily v prirode", Nauka, Moskva, 1977., str. 291.
/40/ Stoiljković D., "Aktuelnost Boškovićeve »Teorije prirodne filozofije svedene na jedan jedini zakon sila koje postoje u prirodi«", Vasiona, **53**, 77-87 (2005)
/41/ Stoiljković D., "Sažimanje materije - Odjeci Boškovićevih shvatanja u teoriji Savić-Kašanin", Vasiona, 53 (4) 178-184 (2005)

/42/ Savić P., Kašanin R., "The behaviour of the materials under high pressures", Serbian academy of sciences and arts, Monographs, Vol. 351, Section for Natural Sciences and Mathematics, No. 29, Beograd, 1962.
/43/ Savić P., "O nastanku rotacije sistema i pojedinih nebeskih tela", Glas SANU, CCXLV, **21**, 37-43 (1961)
/44/ Savić P., "Od atoma do nebeskih tela - poreklo rotacije nebeskih tela", second edition, "Radivoj Ćirpanov", Novi Sad, 1978.
/45/ Stoiljković D., "Dijalektičko-materijalistička osnova teorije Savić-Kašanin o ponašanju materije pri visokim pritiscima i o nastanku rotacije nebeskih tela", Dijalektika, **14**, 137 (1979)
/46/ Savić P., "O »atomistici« R. J. Boškovića", Dijalektika, **32**, 7 (1987)
/47/ Dean J. A., "Lange's Handbook of Chemistry", 12th edition, McGraw-Hill, New York, 1979.
/48/ Filippov L. P., "Podobye svoistv veshchestv", Izdael'stvo Moskovskogo universiteta, Moskva, 1978.
/49/ Stoiljković D., Macanović R., Pošarac D., "The correlation between characteristic volumes of matter - a mathematical model and its physical meaning", J. Serb. Chem. Soc., **60**, 15 (1995)
/50/ Stoiljković D., "Karakteristične zapremine materije - izračunavanje, smisao i značaj", plenary lecture, Zbornik radova sa V Sastanka hemičara Vojvodine, Kikinda, 18.-20.09.1986.
/51/ Syunyaev R. A., "Fizika kosmosa", Sovetskaya entsiklopediya, Moskva, 1986.
/52/ Galaksija, br. 4, 33 (1988)
/53/ Stoiljković D., "Mehanizam i kinetika polimerizacije etilena pri visokom pritisku", doktorska disertacija, Tehnološko-metalurški fakultet, Beograd, 1981.
/54/ Stoiljković D., Jovanović S., "Relations between characteristic volumes of matter", Bull. Soc. Chim., Beograd, **48**, 49-54 (1983)
/55/ Stoiljković D., "Uticaj fizičko–hemijskog stanja etilena na mehanizam i kinetiku polimerizacije po tipu slobodnih radikala pri visokom pritisku", magistarski rad, Tehnološko–metalurški fakultet, Beograd, 1978.
/56/ Hunter E., "The reaction kinetics of ethylene polymerization" in Renfrew A. i Morgan P., eds., "Polythene", Iliffe and Sons, London, 1957.
/57/ Stoiljković D., Jovanović S., "The mechanism of the high pressure free-radical polymerization of ethylene", J. Polym. Sci, Polymer Chem. Ed., **19,** 741-747 (1981)
/58/ Stoiljković D., Jovanović S., "Mechanism of the short chain branching in low density polyethylene", Makromol. Chem., **182**, 2811-2820 (1981)
/59/ Stoiljković D., Jovanović S., "Einfluss der Anderung des Ordnungs-grades des Ethylens mit der Druckerhöchung auf den Verlauf der radikalischen Polymerisation", Angew. Makromol. Chem., **106**, 195-205 (1982)
/60/ Stoiljković D., Jovanović S., "Origin of the molecular structure and properties of low density polyethylene", Brit. Polymer J., **16**, 291-300 (1984)

/61/ Stoiljković D., Jovanović S., "Comments on the articles "Thermisch und UV-photochemisch initiierte Hochdruck-Polymerissation des Ethylens"", Makromol. Chem., **186**, 671-674 (1985)

/62/ Stoiljković D., Jovanović S., "Supermolecular organization and polymerization of compressed ethylene", Acta Polymerica, **39**, 670-676 (1988)

/63/ Stoiljković D., Radičević R., Janković M., "Dependence of the structural parameters and properties of low density polyethylene on the synthesis conditions", J. Serb. Chem. Soc., **64,** 577 (1999)

/64/ Stoiljković D., Pilić B., Radičević R., Bakočević I., Jovanović S., Panić D., Korugić-Karasz Lj., "Polymerization of organized monomers", Hem. ind., **58**, 479 (2004)

/65/ Stoiljković D., Damjanović B., Đorđević J., Špehar D., Jovanović S., "Compressed ethylene phase states and their importance for the production of low density polyethylene", Hem. ind., **60,** 283-286 (2006)

/66/ Radičević R., Stoiljković D., "Samoubrzanje radikalne polimerizacije - šest decenija aktuelan fenomen", Hem. ind., **53**, 336-343 (1999)

/67/ Kargin V. A., Kabanov V. A., "Polimerizatsiya v strukturirovannykh sistemakh", Zh. vses. khim. obshch. im. D. I. Mendeleeva, **9**, 602 (1964)

/68/ Eyring A., Marchi R. P., ""Significant structure theory of liquids", J. Chem. Edu., **40**, 562-572 (1963)

/69/ Sasuga T., Takehisa M., "Radiation-induced polymerization of methyl methacrylate at high pressure", J. Macromol. Chem., **A12**, 1307 (1978); Sasuga T., Takehisa M. "Pressure-volume behaviour of PMMA-MMA coexistence system as polymerized at high pressure", **A12**, 1321 (1978)

/70/ Korolev G. V., Mogilevich M. M., Ily'in A. A., "Assotsiatsiya zhidkikh organicheskikh soedinenii: vliyanie na fizicheskie svoistva i polimerizatsionnye protsessy", Mir, Moskva, 2002.

/71/ Boscovich R., "De viribus vivis", 1745.; Cited according to Martinović I., "The philosophy of science of Ruđer Bošković", Institute of philosophy and theology, Croatian province of the Society of Jesus, Zagreb, 1987., p. 68.

/72/ Korugić Lj., "Nadmolekulska organizacija i polimerizacija metilmetakrilata", doktorska disertacija, Tehnološki fakultet, Novi Sad, 1986.

/73/ Radičević R., Korugić Lj., Stoiljković D., Jovanović S., "Supermolecular organization and characteristic moments of the polymerization of methyl methacrylate", J. Serb. Chem. Soc., **60**, 347-363 (1995)

/74/ "The philosophy of science of Ruđer Bošković", Institute of philosophy and theology, Croatian province of the Society of Jesus, Zagreb, 1987.

/75/ Filozofska istraživanja, 32-33. God. **9**, Sv. 5-6, 1459-1638 (1989)

/76/ Nedeljković D., "Ruđer Bošković u svome vremenu i danas", Kultura, Beograd, 1961.

/77/ Oster M., "Roger Joseph Boscovich als Naturphilosoph", Inaugural-Dissertation, Philosophischen Fakultät, Bonn, Druck von Heinrich Theissing, Cöln, 1909.

/78/ Stoiljković D., "Atrakcija i repulzija - shvatanja Boškovića, Hegela i Engelsa", Filozofska istraživanja, 32-33. God. **9**, Sv.5-6, 1567 (1989)

/79/ Hegel G. V. F., "Nauka logike", Prvi deo, BIGZ, Beograd, 1976., 164-180.
/80/ Hegel G. V. F., "Enciklopedija filozofskih znanosti", Veselin Masleša, Sarajevo, 1965., 114, 138-141, 217-218, 226.
/81/ Engels F., "Dijalektika prirode", Kultura, Beograd, 1970.
/82/ Nonhebel D. C., Walton J. C., "Free-radical chemistry", Cambridge university press, Cambridge, 1974.
/83/ Lederman L., Terezi D., "Božija čestica", Sfinga, Beograd, 1998. (Lederman L., Teresi D., "The God particle", 1993.)
/84/ Paušek-Baždar S., "Kemijski aspekti Boškovićeve teorije", Jugoslavenska akademija znanosti i umjetnosti, Rasprave i građa za povijest znanosti - knjiga 4, Razrada za matematičke, fizičke i tehničke znanosti, Zagreb, 1983.
/85/ Kutleša S., "Prirodno-filozofijski pojmovi Ruđera Boškovića", Hrvatsko filozofsko društvo, Zagreb, 1994.
/86/ Rytter E., Gruen D. M., "Infrared spectra of matrix isolated and solid ethylene. Formation of ethylene dimers", Spectrochimica Acta, 354, 199 (1979)
/87/ Jašo V., Stoiljković D., Radičević R., Bera O., "Kinetic modelling of bulk free radical polymerization of methyl methacrylate", Polymer J., **45**, 631-636 (2013)
/88/ Bera O., Radičević R., Stoiljković D., Jovičić M., Pavličević J., "A new approach for the kinetic modelling of free radical bulk polymerization of styrene", Polymer J., **43**, 826-831 (2011)
/89/ Bera O., Pavličević J., Jovičić M., Stoiljković D., Pilić B., Radičević R., "The influence of nanosilica on styrene free radical polymerization kinetics", Polymer Composites, **33** (2) 262-266 (2012)
/90/ Thomson J. J., "The corpuscular theory of matter", Charles Scribner's Sons, New York, 1907. p. 160)

## Note about author

**Dr. Eng. Dragoslav Stoiljkovich** was born in 1947 in Belgrade (Serbia). He graduated in 1971, finished master science degree in 1978 and Ph.D. in 1981 at University of Belgrade, Faculty of Technology and Metallurgy. Master's theses and doctoral dissertations were devoted to the polymerization of ethylene at high pressure.

From 1971 up to 1981 was employed in the Chemical Industry "Panchevo" in different positions, from plant engineer up to manger of Factory for production of low density polyethylene. Since 1981 up to retirement in 2012 employed at University of Novi Sad, Serbia, at the Faculty of Technology, Department of Materials engineering, as regular professor of the production and processing of polymeric materials, as well as for the case "The methodology of scientific research work". In the period since 1990 up to 1996 he was employed in the Serbian oil and gas industry (NIS) where he worked on the restructuring of the petroleum, petrochemical and polymer industry in Serbia.

He participated (as a manager or collaborator) in 23 scientific research projects on the effects of supra-molecular organization of the monomer on the mechanism and kinetics of polymerization, as well as the processing and application of different polymers. He published 105 articles and 172 reports, of which 60 articles and 57 reports were dedicated to Roger Boscovich, his Theory and its application.

Member of the Presidency of the Association of Chemical Engineers of Serbia, member of the Presidency of the Serbian Chemical Society (SHD) and President of the Section for materials within the SHD-Chemical Society of Province Vojvodina.

CIP - Cataloguing in Publication
National Library of Serbia, Belgrade
5:929 Boscovich R.
**Stoiljkovich. Dragoslav 1947 -**
Ruđer Bošković – utemeljivač savremene
nauke / Dragoslav Stoiljković - Petnica :
Petnica Science Center, 2010 (Valjevo :
Valjevoprint). 82 p. : ilustr. 24 cm.
(Petnica notebooks, ISSN 0354-1428, No. 65).

Circulation 400 -Note about author: p. 81 -
Bibliography: p. 75-79.
ISBN 978-86-7861-043-1
A) Bošković, Ruđer, Roger (1711-1787)
COBISS.SR ID 174596876

www.ingramcontent.com/pod-product-compliance
Lightning Source LLC
Chambersburg PA
CBHW080946170526
45158CB00008B/2393